Nutrition and Gene Interaction

Nutrition and Gene Interaction

Guest Editor

Marica Bakovic

MDPI

Basel • Beijing • Wuhan • Barcelona • Belgrade • Novi Sad • Cluj • Manchester

Guest Editor
Marica Bakovic
Department of Human
Health and Nutritional
Sciences
University of Guelph
Guelph
Canada

Editorial Office
MDPI AG
Grosspeteranlage 5
4052 Basel, Switzerland

This is a reprint of the Special Issue, published open access by the journal *Nutrients* (ISSN 2072-6643), freely accessible at: https://www.mdpi.com/journal/nutrients/special_issues/JN7IDZZI11.

For citation purposes, cite each article independently as indicated on the article page online and as indicated below:

Lastname, A.A.; Lastname, B.B. Article Title. *Journal Name* **Year**, *Volume Number*, Page Range.

ISBN 978-3-7258-4703-7 (Hbk)
ISBN 978-3-7258-4704-4 (PDF)
https://doi.org/10.3390/books978-3-7258-4704-4

Contents

About the Editor

Marica Bakovic

Marica Bakovic is an internationally recognized researcher in membrane biogenesis, nutrient transport, and natural health product characterization. Her area of expertise includes the genomic and non-genomic regulation of metabolism, with specific emphasis on phospholipids and methyl-group donors. Her laboratory has established strong molecular links between membrane lipid impairments and energy metabolism and continues to pursue studies on the structure and function of a key regulatory enzyme (Pcyt2) in membrane phospholipid synthesis, a process linked to several chronic metabolic diseases including obesity, hypertriglyceridemia, insulin resistance, and nonalcoholic steatohepatitis (NASH). Recently, her group discovered a new childhood-onset neurodegenerative disease (CONATOC) caused by mutations in choline transporter-like protein 1 (Slc44A1/CTL1). Her research in this area has been critical in establishing the genetic and molecular links between disease progression and choline function in brain metabolism, mitochondria oxidation, and regulation by methylation. In addition to her cutting-edge basic mechanistic research on lipid metabolism, Dr. Bakovic is among the first scientists to investigate gene–diet interactions related to personalized nutrition, and to incorporate this novel thought process into undergraduate and graduate curricula in nutritional genomics. Dr. Bakovic's highly productive and impactful research program has produced over 350 contributions, including 105 articles in referred journals, 13 referred books and book chapters, and 8 patents and trademarks. She has edited several books and journals, including *Brain Choline Transport and Function* and *Functional Foods, Nutraceuticals*, and *Degenerative Disease Prevention*. She has received numerous awards, including the Senior Investigator Award and Simon-Pierre Noel Award from the Canadian Lipid and Lipoprotein Society-Conference and the Research Excellence Award from the Government of Ontario.

Preface

Over the past 15 years, the field of nutrigenomics has significantly advanced our understanding of the intricate relationship between genes, nutrition, and disease. This burgeoning area of research has demonstrated that nutrition is not merely about preventing deficiencies, it also holds immense potential for preventing and treating chronic illnesses such as diabetes, cancer, and neurodegenerative disorders. By elucidating how nutrients modulate cellular signaling pathways, diet can be harnessed as a first line of defense against these chronic conditions.

The importance of characterizing genes and analyzing their expression under various nutritional conditions cannot be overstated. This research not only informs dietary recommendations at the individual level but also guides the development of public dietary guidelines aimed at reducing the prevalence of chronic illnesses in the population.

This Special Issue compiles recent research on the impact of nutrients on genetic expression, employing methods such as transcriptomics, proteomics, and metabolomics. It presents nine studies that explore gene–diet interactions across a range of health outcomes, highlighting both the promise and complexity of nutrigenomics. The key findings presented in this Special Issue include the role of fermented foods in reducing the risk of non-alcoholic fatty liver disease, the influence of genetics on dietary needs and bone health, and the impact of early-life nutrition on long-term gene expression. The Special Issue also addresses the genetic regulation of liver, cardiovascular, and thyroid diseases, and the potential for personalized nutrition interventions to mitigate disease risk.

Overall, this reprint advocates for precision nutrition, leveraging genetic and epigenetic insights to create personalized dietary strategies for better health outcomes. As nutrigenomics continues to evolve, future research must validate these findings through clinical trials and explore how best to integrate personalized nutrition into public health strategies.

Marica Bakovic
Guest Editor

nutrients

MDPI

Editorial

Advancing Personalized Nutrition Through Genetic Nutritional Insights

Michaela Marie St Germain, Amirreza Mortazavi and Marica Bakovic *

Marica Bakovic's Laboratory, Human Health and Nutritional Sciences Department, University of Guelph, 50 Stone Rd. E, Guelph, ON N1G 2W1, Canada; mstg0001@student.monash.edu (M.M.S.G.); amortaza@uoguelph.ca (A.M.)
* Correspondence: mbakovic@uoguelph.ca

Metabolic health is shaped by the dynamic interplay between genetic predispositions and dietary habits. As our understanding of this relationship deepens, we move closer to a nutritional framework that is not only evidence-based, but also personalized to each individual's genetic profile. This Special Issue of *Nutrients* presents nine studies that explore gene–diet interactions across a range of health outcomes, highlighting both the promise and the complexity of nutrigenomics.

Given its strong links to both diet and genetic susceptibility, metabolic disease is a particularly relevant focus within nutrigenomics. This was addressed in the work of Oh et al. (2023) [contribution 1], which found that variations in the PNPLA3 gene significantly modulate non-alcoholic fatty liver disease (NAFLD) susceptibility, and that this risk was reduced in individuals with increased kimchi intake. This points to a potential role for fermented foods in the prevention of NAFLD in individuals with genetic predispositions. Such findings exemplify how personalized dietary guidance, shaped by genetic screening, could offer substantial health benefits by mitigating genetic risks through specific nutritional strategies.

Expanding beyond NAFLD, Park (2023) [contribution 2] examined height-related polygenic variants, linking them to metabolic syndrome risk. Interestingly, taller individuals with specific genetic profiles had a reduced risk of metabolic syndrome, especially when consuming diets lower in rice and higher in energy. This underscores the influence of genetics on dietary needs and raises important considerations for culturally appropriate dietary recommendations. The study also investigated functional gene–diet interactions, showing that the GDF5_rs224331 variant alters how its protein product binds hydrolysable tannins—compounds known to support bone health. These findings illustrate how genetic variation shapes nutrient responses and can inform personalized strategies for managing chronic conditions like osteoporosis.

In addition to osteoporosis, this Special Issue includes novel insights into the genetic regulation of liver, cardiovascular, and thyroid disease. Ghare et al. (2023) [contribution 3] demonstrated that tributyrin, a butyrate precursor found in dairy and legumes, reduced alcohol-induced liver inflammation by reversing harmful epigenetic changes. This work not only highlights a therapeutic avenue for alcohol-associated liver disease, but also exemplifies how nutrient-derived compounds can directly modulate gene expression. Additionally, the study by Garrido-Sanchez et al. (2024) [contribution 4] sheds light on the connection between single nucleotide variants in the ABCG8 gene and phytosterol metabolism, which has associations with parenteral nutrition-associated liver disease and cardiovascular disease risk. Furthermore, Kim and Park (2023) [contribution 5] revealed significant interactions between polygenic variants related to hypothyroidism and lifestyle factors, including dietary habits and smoking. Such genetic–diet interactions offer promising pathways for personalized nutrition interventions aimed at reducing the incidence and severity of hypothyroidism, further demonstrating the vast potential for diet to modulate genetic risk.

Beyond disease risk, genetics can also influence dietary behaviors. Franzago et al. [contribution 6] (2023) found that variants in the CD36 gene—implicated in fat taste perception—were associated with differences in anthropometric and metabolic outcomes among individuals with diabetes or dysglycemia. These findings raise the possibility that sensory genetics could help tailor dietary counseling to improve adherence and outcomes.

Early-life nutrition and epigenetic mechanisms add another critical dimension to gene–diet interactions, as nutritional exposures during sensitive developmental periods can influence long-term gene expression through processes like DNA methylation. Patel et al. (2023) [contribution 7] highlighted the significance of dietary methyl donors such as folate, demonstrating their impact on methylation patterns of obesity-related genes like NRF1, FTO, and LEPR in children. This study illustrates the promise of early nutritional intervention but also the need for future research that accounts for intersecting biological and social determinants of health.

The study by Buchanan et al. (2024) [contribution 8] also investigated how nutrition during developmental periods can impact future disease using a mouse model of lifelong dietary exposure to n-3 polyunsaturated fatty acids (PUFAs). Their findings suggest that early n-3 PUFA intake, particularly during puberty, promotes mammary gland differentiation and reduces pro-carcinogenic gene expression patterns, pointing to the preventive potential of diet during sensitive developmental windows.

While many of the studies in this Special Issue report promising associations between genetics and dietary responses, it is equally important to acknowledge null findings that refine our understanding of where gene–diet interactions may or may not be significant. For instance, Górczyńska-Kosiorz et al. (2024) [contribution 9] finds no significant relationship between the obesity-related FTO gene, dietary patterns, and metabolic syndrome in a subset of young, healthy Polish men. However, the absence of significant findings itself reinforces a key point that genetic risk does not guarantee disease. Factors such as culture, environment, and protective behaviors can buffer genetic predisposition.

Overall, this Special Issue of *Nutrients* covers a wide range of topics in nutrigenetics and nutrigenomics; however, the overall message is clear. Diet and genetics are interdependent, and their impacts on chronic disease cannot be considered separately. These integrated findings from recent studies strongly advocate a shift towards precision nutrition, which leverages comprehensive genetic and epigenetic insights. In order to facilitate this transition, it is crucial to take the findings from observational and animal-based studies and transitions to clinical research. For example, based on the success of tributyrin supplementation in mouse models of alcohol-related liver disease, it could be reasonable to move toward a randomized controlled trial in humans. Furthermore, it would be interesting to explore the mechanisms behind gene and diet associations to gain a deeper understanding of the physiological impacts of nutrients. Namely, it would be interesting to explore potential causation and correlation between kimchi consumption and reduced risk of NAFLD in people with genetic predispositions, as this might be extrapolated to other fermented foods, as well as other chronic disease.

Finally, with all of this in mind, the feasibility of personalized, gene-based dietary counseling must be considered. This will require further research into the ethical, economic, and logistical challenges of implementation, including patient privacy, socioeconomic disparities, healthcare infrastructure, and the translation of genetic data into clinical practice. As nutrigenomics continues to evolve, future research must not only validate gene–diet interactions through clinical trials, but also investigate how best to integrate these insights into equitable, accessible, and effective public health strategies. The question now is not whether personalized nutrition is possible, but how we can responsibly and rigorously build the evidence to support its widespread, real-world application.

Conflicts of Interest: The authors declare no conflicts of interest.

List of Contributions:

1. Oh, S.; Lee, J.; Chun, S.; Choi, J.-E.; Kim, M.N.; Chon, Y.E.; Ha, Y.; Hwang, S.-G.; Choi, S.-W.; Hong, K.-W. Interaction between the PNPLA3 Gene and Nutritional Factors on NAFLD Development: The Korean Genome and Epidemiology Study. *Nutrients* **2023**, *15*, 152. https://doi.org/10.3390/nu15010152.
2. Park, S. Height-Related Polygenic Variants Are Associated with Metabolic Syndrome Risk and Interact with Energy Intake and a Rice-Main Diet to Influence Height in KoGES. *Nutrients* **2023**, *15*, 1764. https://doi.org/10.3390/nu15071764.
3. Ghare, S.S.; Charpentier, B.T.; Ghooray, D.T.; Zhang, J.; Vadhanam, M.V.; Reddy, S.; Joshi-Barve, S.; McClain, C.J.; Barve, S.S. Tributyrin Mitigates Ethanol-Induced Lysine Acetylation of Histone-H3 and p65-NFκB Downregulating CCL2 Expression and Consequent Liver Inflammation and Injury. *Nutrients* **2023**, *15*, 4397. https://doi.org/10.3390/nu15204397.
4. Garrido-Sanchez, L.; Leiva-Badosa, E.; Llop-Talaveron, J.; Pintó-Sala, X.; Lozano-Andreu, T.; Corbella-Inglés, E.; Alia-Ramos, P.; Arias-Barquet, L.; Ramon-Torrel, J.M.; Badía-Tahull, M.B. Blood Phytosterol Concentration and Genetic Variant Associations in a Sample Population. *Nutrients* **2024**, *16*, 1067. https://doi.org/10.3390/nu16071067.
5. Kim, D.S.; Park, S. Interactions between Polygenetic Variants and Lifestyle Factors in Hypothyroidism: A Hospital-Based Cohort Study. *Nutrients* **2023**, *15*, 3850. https://doi.org/10.3390/nu15173850.
6. Franzago, M.; Borrelli, P.; Di Nicola, M.; Stuppia, L.; Vitacolonna, E. Genetic Variants in CD36 Involved in Fat Taste Perception: Association with Anthropometric and Clinical Parameters in Overweight and Obese Subjects Affected by Type 2 Diabetes or Dysglycemia—A Pilot Study. *Nutrients* **2023**, *15*, 4656. https://doi.org/10.3390/nu15214656.
7. Patel, P.; Selvaraju, V.; Babu, J.R.; Geetha, T. Association of the DNA Methylation of Obesity-Related Genes with the Dietary Nutrient Intake in Children. *Nutrients* **2023**, *15*, 2840. https://doi.org/10.3390/nu15132840.
8. Buchanan, C.D.C.; Ashraf, R.; Hillyer, L.M.; Tu, W.; Kang, J.X.; Subedi, S.; Ma, D.W.L. RNA-Seq Analysis of Pubertal Mammary Epithelial Cells Reveals Novel n-3 Polyunsaturated Fatty Acid Transcriptomic Changes in the fat-1 Mouse Model. *Nutrients* **2024**, *16*, 3925. https://doi.org/10.3390/nu16223925.
9. Górczyńska-Kosiorz, S.; Lejawa, M.; Goławski, M.; Tomaszewska, A.; Fronczek, M.; Maksym, B.; Banach, M.; Osadnik, T. The Impact of Haplotypes of the FTO Gene, Lifestyle, and Dietary Patterns on BMI and Metabolic Syndrome in Polish Young Adult Men. *Nutrients* **2024**, *16*, 1615. https://doi.org/10.3390/nu16111615.

nutrients

MDPI

Communication

RNA-Seq Analysis of Pubertal Mammary Epithelial Cells Reveals Novel *n*-3 Polyunsaturated Fatty Acid Transcriptomic Changes in the *fat-1* Mouse Model

Connor D. C. Buchanan [1,†], Rahbika Ashraf [1,†], Lyn M. Hillyer [1], Wangshu Tu [2], Jing X. Kang [3,4], Sanjeena Subedi [2] and David W. L. Ma [1,*]

1 Department of Human Health and Nutritional Sciences, University of Guelph, Guelph, ON N1G 2W1, Canada; cbucha03@uoguelph.ca (C.D.C.B.); rashraf@uoguelph.ca (R.A.); lhillyer@uoguelph.ca (L.M.H.)
2 School of Mathematics and Statistics, Carleton University, Ottawa, ON K1S 5B6, Canada; wangshutu@cunet.carleton.ca (W.T.); sanjeenadang@cunet.carleton.ca (S.S.)
3 Laboratory for Lipid Medicine and Technology, Department of Medicine, Massachusetts General Hospital and Harvard Medical School, Boston, MA 02129, USA; jxkang@oghi.org
4 Omega-3 and Global Health Institute, Boston, MA 02129, USA
* Correspondence: davidma@uoguelph.ca
† These authors contributed equally to this work.

Abstract: Background: The early exposure of nutrients during pubertal mammary gland development may reduce the risk of developing breast cancer later in life. Anticancer *n*-3 polyunsaturated fatty acids (*n*-3 PUFA) are shown to modulate pubertal mammary gland development; however, the mechanisms of action remain unclear. Prior work focused on effects at the whole tissue level, and little is known at the cellular level, such as at the level of mammary epithelial cells (MECs), which are implicated in cancer development. Methods: This pilot study examined the effects of lifelong *n*-3 PUFA exposure on the transcriptome by RNA-Seq in the isolated MECs of pubertal (6–8-week-old) female *fat-1* transgenic mice capable of de novo *n*-3 PUFA synthesis. *edgeR* and *DESeq2* were used separately for the differential expression analysis of RNA sequencing data followed by the Benjamani–Hochberg procedure for multiple testing correction. Results: Nine genes were found concordant and significantly different ($p \leq 0.05$) by both the DESeq2 and edgeR methods. These genes were associated with multiple pathways, suggesting that *n*-3 PUFA stimulates estrogen-related signaling (*Mlltl0*, *Galr3*, and *Nrip1*) and a glycolytic profile (*Soga1*, *Pdpr*, and *Uso1*) while offering protective effects for immune and DNA damage responses (*Glpd1*, *Garre1*, and *Rpa1*) in MECs during puberty. Conclusions: This pilot study highlights the utility of RNA-Seq to better understanding the mechanistic effects of specific nutrients such as *n*-3 PUFA in a cell-specific manner. Thus, further studies are warranted to investigate the cell-specific mechanisms by which *n*-3 PUFA influences pubertal mammary gland development and breast cancer risk later in life.

Keywords: *n*-3 polyunsaturated fatty acids; eicosapentaenoic acid; docosahexaenoic acid; alpha-linolenic acid; mammary epithelial cells; RNA-Seq

1. Introduction

Breast cancer (BC) is a leading cancer diagnosed worldwide, with 2.3 million cases reported in 2022 [1] and is associated with modifiable risk factors [2]. At least 30% of cancer cases have been associated with lifestyle and dietary habits, including the intake of different types of dietary fats [2]. A number of studies have demonstrated that increasing intakes of *n*-3 polyunsaturated fatty acids (*n*-3 PUFA), which are found in fish and marine oils, may have protective effects against BC [3–5]. In rodent models of BC, lifelong intake of *n*-3 PUFA has been shown to reduce tumor numbers, which may be attributed to its effects on mammary gland development [4,5]. During puberty, female rodents experience extensive growth of the mammary gland driven by rapidly proliferating epithelial cells which form club-like structures known as terminal end buds (TEBs) that contain tumor initiation sites [6,7]. These highly proliferating cells are also responsible for the maturation of the ductal network and are susceptible to cancer initiation [7,8]. The intake of *n*-3 PUFA has been reported to affect pubertal mammary gland development by delaying puberty onset and reducing the number of TEBs, as well as by affecting mammary epithelial cell fate [9,10]. These findings highlight a number of effects of *n*-3 PUFA, but the precise mechanisms of action remain elusive.

To date, most studies have taken a whole-mammary-gland approach. Thus, there remains much to learn at the cellular level concerning specific cell types within the mammary gland. Recent work using cutting-edge bulk RNA sequencing (RNA-Seq) in the analysis of thousands of transcripts in tandem on isolated mammary epithelial cells has revealed cell-specific changes during the course of mammary gland development in goats [11] and in experimental mouse studies [12]. The effects of diet and *n*-3 PUFA at the cellular level have provided powerful insights into isolated mouse colonocytes [13]. However, potential novel insights into the effects of *n*-3 PUFA at the cellular level by RNA-Seq have yet to be investigated.

The *fat-1* mouse model has previously been used to study the effects of lifelong *n*-3 PUFA exposure and its mechanisms of action in many diseases, including BC [5,6], and in mammary gland development [9,10]. The *fat-1* transgene from the roundworm *Caenorhabditis elegans* encodes for *n*-3 desaturase, enabling endogenous whole-body production of *n*-3 PUFA from *n*-6 PUFA, including in the mammary gland [9,10]. This genetic approach makes the study of *n*-3 PUFA and its causal effects possible using the transgenic *fat-1* mouse model while reducing confounding effects from dietary feeding [14], thereby allowing for a precise understanding of the mechanisms of action of lifelong *n*-3 PUFA intake. Thus, this pilot study investigates the transcriptomic effects of lifelong *n*-3 PUFA exposure in mammary epithelial cells isolated from female transgenic *fat-1* mice during puberty using bulk RNA-Seq.

2. Materials and Methods

2.1. Animals, Diets, and Phenotyping

Transgenic *fat-1* mice acquired from Dr. Kang (Harvard Medical School) were used to develop an in-house breeding colony on an FVB background at the University of Guelph, as previously described [5,9,10]. The mice were fed a modified AIN93G diet (Research Diets Inc., New Brunswick, NJ, USA) containing 10% fat (*w*/*w*) from safflower oil, rich in *n*-6 PUFA, providing 22% of the mouse's total daily energy intake requirements, as previously described [10]. In brief, safflower oil contains 70% of the essential *n*-6 PUFA linoleic acid, which transgenic *fat-1* mice will endogenously utilize to synthesize individual *n*-3 PUFA in tissues, including in the mammary gland [9,10]. Female offspring were weaned and phenotyped at three weeks of age and maintained on their parental diets until termination at 6 to 8 weeks. In rodents, puberty mammary gland development has typically been described as occurring from the age of 4 weeks up to 10 weeks [6]. To minimize the number of animals, mice mammary glands were pooled to obtain a sufficient quantity of RNA for bulk RNA Seq from epithelial cells ($n = 3$). Each sample ($n = 1$) represents the pooling of inguinal mammary glands from 2 to 3 mice. Overall, there were $n = 3$ pooled samples each for wild-type (WT) and *fat-1* mice.

2.2. Euthanization, Tissue Collection, Epithelial Cell Isolation, and RNA Extraction

The mice were terminated and the right and left 4th and 5th mammary glands (MGs) were excised for epithelial cell isolation, as previously described [10]. A total of 6 to 8 mice were pooled for each analysis of epithelial cells ($n = 3$); lymph nodes were removed, and mammary epithelial cells were isolated using the Prater method, as previously described [10,15]. In brief, finely minced MGs were digested in collagenase/hyaluronidase (StemCell, cat # 07912) for 18 h at 37 °C to allow for tissue dissociation. The cells were washed in Hank's balanced salt solution (Sigma, cat # H6648) and treated with ammonium chloride (Sigma, cat # A9434), Trypsin/EDTA (Sigma, cat # T4049), dispase (StemCell, cat # 07913), and DNAase 1 (Sigma, cat # D5025) to release the epithelial cells. Five million cells were used for total RNA extraction. RNA was extracted using the Purelink RNA Mini Kit (Thermo Fisher Scientific, cat # 12183018A) following the kit instructions. The purity of the RNA was assessed using the Agilent bioanalyzer 2100 where all samples had an RNA integrity number (RIN) greater than 9 (out of a 10-point scale). RNA samples were stored at −80 °C for later analysis.

2.3. RNA Sequencing and RNA Sequencing Analysis

The samples were sent to the Centre for Applied Genomics in the Hospital for Sick Children (Toronto, ON, Canada) for RNA sequencing. RNA sequencing was performed using the Illumina platform (paired-end reads of a 100 bp sequence), and mapped to the reference genome GRCM38 using HISAT2, as previously described [16]. The results were sent to Carleton University (Subedi) for further analysis. The pilot dataset was analyzed following the protocol from Pertea et al. [17] (the "new Tuxedo" package). The reads from the samples were mapped to the reference genome GRCM38 using HISAT2 [18]. All six samples had an overall alignment rate greater than 97%. The alignments were then passed to String Tie [17] for transcript assembly and quantification. The transcript abundance matrix from String Tie was then used for differential expression analysis using two different approaches: *edgeR* [19] and *DESeq2* [20]. The number of genes that had a p-value less than or equal to 0.05 were 606 and 1154 for *edgeR* and *DESeq2*, respectively (Supplementary Tables S1 and S2). Furthermore, the Benjamani–Hochberg procedure was used to adjust for multiple hypothesis testing [21] separately in the *edgeR* and *DESeq2* analyses. Using a threshold of 0.05 for the false discovery rate (FDR), 31 and 101 genes were identified as differentially expressed using *edgeR* and *DESeq2*, respectively (Supplementary Tables S3 and S4). Only genes with a known nomenclature are reported in this paper.

3. Results

Significant differences in the total RNA expression of 11 genes using *edgeR* (Table 1) and 45 genes using *DESeq2* (Table 2) were identified from the isolated mammary epithelial cells of *fat-1* mice compared to WT mice (adjusted $p \leq 0.05$). Analysis using *edgeR* showed that there were five downregulated and six upregulated genes compared to WT mice (Table 1). For *DESeq2* analysis, there were 18 downregulated and 27 upregulated genes in the *fat-1* mice compared to WT mice (Table 2). Of these genes, nine were found to be significant across both the *edgeR* and *DESeq2* methods (adjusted $p \leq 0.05$), where five genes were downregulated, and four genes were upregulated (Figure 1). The log2 fold changes for nine overlapping genes ranged from 1.14 to 7 (Figure 1).

A comparison of key pathways revealed distinct gene expression patterns between *fat-1* and WT mice. This study found that genes associated with estrogen-related signaling were stimulated with the downregulation of *mixed-lineage leukemia (Mllt10/AF10)* and a gene of *galanin receptor 3* (*Galr3*) and the upregulation of an essential marker of mammary gland development *nuclear receptor interacting protein 1* (*Nrip1/RIP140*) in *fat-1* mice compared to WT mice (Figure 1). The study also found that the changes in gene expression involved in glycolysis were upregulated, as indicated by the upregulation of *suppressor of glucose (Soga1)*, which is a negative regulator of gluconeogenesis, and *uso1 vesicle docking factor (Uso1)*, which is a regulator of insulin stimulus and the downregulation of *pyruvate dehydrogenase phosphatase regulatory subunit (Pdpr)* in *fat-1* mice compared with WT mice. On the other hand, there were effects in genes regulating immune and DNA damage responses with an upregulation in *granule associated rac and rhog effector* 1 (*Garre1*) and *replication protein a1* (*Rpa1*) along with a downregulation in *glycosylphosphatidylinositol specific phospholipase d1* (*Glpd1*) in *fat-1* mice compared to WT mice.

Table 1. Effects of lifelong *n*-3 polyunsaturated fatty acid exposure on gene expression in isolated pubertal mammary epithelial cells between *fat-1* and WT mice assessed by *edgeR*.

Gene	Log_2 Fold Change	Adjusted *p*-Value	Function(s):	NIH National Library of Medicine Gene ID
		edgeR		
Galr3 *, galanin receptor 3	−7.00	<0.0001	Positive regulation of transcription by RNA polymerase II, G-protein-coupled receptor signaling pathway, peptide hormone binding activity, estrogen response marker [22]	14429
Garre1, granule associated rac and rhog effector 1	−1.71	<0.0001	Enables CCR4-Not complex binding activity and Rac signaling production	233103
Gpld1 *, glycosylphosphatidylinositol-specific phospholipase d1	−5.18	<0.0001	Enables phospholipase D activity, sodium channel regulator, positive regulator of HDL particle clearance, positive regulation of insulin secretion in response to glucose, regulation of triglyceride metabolic process, immune cell recruitment [23]	14756
Mllt10 (AF10) *, mixed-lineage leukemia	−2.19	<0.0001	Enables histone and nucleosome binding, positive regulation of transcription by RNA polymerase II, H3K79 methylation regulation by DOT1L [24]	17354
Nrip1 (RIP140) *, nuclear receptor interacting protein 1	3.58	<0.0001	Negative regulator of transcription by RNA polymerase II, Circadian rhythm, lipid storage, histone deacetylase complex, retinoid X receptor binding activity. Role in estrogen signaling and mammary gland development [25]	268903
Pdpr *, pyruvate dehydrogenase phosphatase regulatory subunit	−3.08	<0.0001	Enables oxidoreductase activity	319518
Rpa1 *, replication protein a1	1.14	<0.0001	Enables chromatin binding activity, DNA repair, DNA replication, DNA recombination [26]	68275
Slc24a2, solute carrier family 24 member 2	3.72	<0.0001	Enables calcium channel activity	76376
Soga1 *, suppressor of glucose	2.53	<0.0001	Insulin receptor signaling, macronutrient metabolism, negative regulation of gluconeogenesis, regulation of autophagy [27]	N/A
Terc, telomerase rna component	2.80	<0.0001	Involved in telomere maintenance and chromosomal repair	21748
Uso1, uso1 vesicle docking factor	2.10	<0.0001	Regulation of insulin stimulus, secretory granule localization, small GTPase signal transduction, intracellular protein transport, marker of glucose utilization [28]	56041

Data are log_2 fold change. Change in total RNA expression within isolated total mammary epithelial cells of transgenic *fat-1* mice, which genetically reflects lifelong *n*-3 polyunsaturated fatty acid (*n*-3 PUFA) exposure [10], compared to female wildtype (WT) mice on a *n*-6 polyunsaturated fatty acid (*n*-6 PUFA)-rich diet aged 6 to 8 weeks. Three pooled samples from *fat-1* and WT mice underwent RNA sequencing, performed using the Illumina platform (paired-end reads of a 100 bp sequence), and mapped to reference genome GRCM38 using HISAT2 [18]. Differential expression analysis was performed using *edgeR* followed by multiple testing correction using the Benjamani–Hochberg procedure to control for the false discovery rate (FDR) at 5% [21]. Using a threshold of an FDR-adjusted *p*-value of 0.05 revealed 31 genes that were differentially expressed by *edgeR*. Of these, 11 genes with an known nomenclature are reported in this Table. * Genes that were identified as differentially expressed compared to the WT using the edgeR method after Benjamani–Hocheburgh adjustment for multiple hypothesis testing. Gene functions were found using Gene ID description from the NIH National Library of Medicine on 4 November 2023. References were provided for additional gene functions not listed in the NIH National Library of Medicine.

Table 2. Effects of lifelong *n*-3 polyunsaturated fatty acid exposure on gene expression in isolated pubertal mammary epithelial cells between *fat-1* and WT mice assessed by *DESeq2*.

Gene	Log_2 Fold Change	Adjusted *p*-Value	Function(s):	NIH National Library of Medicine Gene ID
		DESeq2		
Arfip1, adp ribosylation factor interacting protein 1	−1.33	<0.0001	Enables phosphatidylinositol-4-phosphate binding activity; negative regulator of retrograde transport	99889
Arl4a, adp ribosylation factor-like gtpase 4	1.10	<0.0001	Enables GTP binding activity	11861
B3gat3, beta-1,3-glucuronyltransferase 3	1.79	0.01	Enables galactosylgalactosylxylosylprotein 3-beta-glucuronosyltransferase activity and protein phosphatase activator activity	72727
Baz2b, bromodomain adjacent to zinc finger domain 2b	−0.84	0.001	Enables DNA binding activity; Chromatin remodeling	407823
Chchd2-ps, coiled-coil-helix-coiled-coil-helix domain containing 2	1.61	0.046	N/A	433806
Clpp, caseinolytic mitochondrial matrix peptidase proteolytic subunit	1.10	0.001	Enables ATP-dependent peptidase; membrane protein proteolysis and protein quality control	53895
Csf2rb2, colony stimulating factor 2 receptor subunit beta	−2.96	0.036	Enables cytokine receptor activity; cytokine-mediated signaling	12984
Cspp1, centrosome and spindle pole-associated protein 1	−0.48	0.0004	Positive regulator of cytokinesis	211660
Dhcr7, 7-dehydrocholesterol reductase	−4.23	0.03	Enables 7-dehydrocholesterol reductase activity (lipid metabolism)	13360
Dohh, deoxyhypusine hydroxylase	1.16	0.0001	Enables deoxyhypusine monooxygenase activity and iron ion binding activity; acts upstream of or within peptidyl-lysine modification to peptidyl-hypusine	102115
Galr3 *, galanin receptor 3	−6.44	0.002	Positive regulation of RNA polymerase II; G-protein-coupled receptor signaling pathway; peptide hormone binding activity; estrogen response marker [22]	14429
Garre1 *, granule-associated rac and rhog effector 1	−1.72	<0.0001	Enables CCR4-Not complex binding activity; Rac signaling	233103
Gigyf1, grb10 interacting gyf protein 1	−0.82	<0.0001	Involved in insulin-like growth factor receptor signaling	57330
Glg1, golgi apparatus protein 1	−0.57	<0.0001	Enables cell surface interactions; fibroblast growth factor binding activity; negative regulator of beta receptor signaling pathway	20340

Table 2. *Cont.*

Gene	Log_2 Fold Change	Adjusted *p*-Value	Function(s):	NIH National Library of Medicine Gene ID
Gpatch8, g-patch domain containing 8	1.42	0.0004	Enables metal ion activity and nucleic acid binding activity	237943
Gpld1 *, glycosylphosphatidylinositol-specific phospholipase d1	−5.20	<0.0001	Enables phospholipase D activity; sodium channel regulator; positive regulator of HDL particle clearance; positive regulation of insulin secretion in response to glucose; regulation of triglyceride metabolic process, immune cell recruitment [23]	14756
Hnrnpl, heterogeneous nuclear ribonucleoprotein l	0.51	0.003	Enables mRNA-binding protein activity; regulator of alternative mRNA splicing	15388
Map3k4, mitogen-activated protein kinase 4	5.05	0.0003	Enables protein kinase activity; positive regulator of JUN kinase activity	26407
Med8, mediator complex subunit 8	0.67	<0.0001	Enables RNA polymerase II cis-regulator region DNA binding activity	80509
Mllt10 (AF10) *, mixed-lineage leukemia	−2.19	0.002	Enables histone and nucleosome binding; positive regulator of transcription by RNA polymerase II, regulating H3K79 methylation by DOT1L [24]	17354
Morc4, morc family cw-type zinc finger 4	0.90	0.007	Enables methylated histone binding activity	75746
Nrip1 (RIP140) *, nuclear receptor interacting protein 1	3.58	<0.0001	Negative regulator of transcription by RNA polymerase II, Circadian rhythm, lipid storage, histone deacetylase complex, retinoid X receptor binding activity Role in estrogen signaling and mammary gland development [25]	268903
Pbxip1, pbx homeobox interacting protein 1	1.80	0.02	Enables transcription coactivator activity, histone H3-K56 acetylation, extracellular matrix organization	229534
Pcdha9, protocadherin alpha 9	6.23	0.011	Enables calcium binding activity; involved in cell adhesion	192161
Pdpr *, pyruvate dehydrogenase phosphatase regulatory subunit	−3.08	0.0002	Enables oxidoreductase activity	319518
Per1, period circadian regulator 1	−0.69	0.008	Involved in circadian rhythm	18626
Phc3, polyhomeotic homolog 3	−2.06	0.03	Negative regulator of transcription, PcG protein complex, chromatin binding activity, histone binding activity	241915
Plin3, perilipin 3	3.67	0.013	Involved in lipid storage and positive regulator of sequestering of triglyceride	66905
Pnp, purine nucleoside phosphorylase	−1.64	0.023	Enables guanosine phosphorylase activity and purine nucleoside phosphorylase activity	18950
Prdm11, pr/set domain 11	−1.05	0.042	Enables chromatin binding activity; negative regulator of cell growth	100042784
Rapgef3, rap guanine nucleotide exchange factor 3	−0.38	0.033	Enables guanyl-nucleotide exchange factor activity.	223864

Table 2. *Cont.*

Gene	Log_2 Fold Change	Adjusted *p*-Value	Function(s):	NIH National Library of Medicine Gene ID
Rbm26, rna binding motif protein 26	0.95	0.0003	Enables RNA binding activity; involved in mRNA processing	74213
Rn7s1, 7s rna	4.30	0.002	N/A	103948
Rn7s6, 7s rna- 6	5.18	0.001	N/A	109568
Rpa1 *, replication protein a1	1.14	<0.0001	Enables chromatin binding activity, DNA repair, DNA replication, DNA recombination [26]	68275
Rrp8, ribosomal rna processing 8	1.93	0.036	Enables methylated histone binding activity; involved in cellular response to glucose starvation, intrinsic apoptotic signaling pathway by p53 class mediator, chromosome organization	101867
Sfr1, swi5-dependent homologous recombination repair protein 1	0.56	0.007	Enables nuclear receptor coactivator activity, double-stand break repair, cell cycle, DNA repair	67788
Soga1 *, suppressor of glucose	2.53	0.008	Insulin receptor signaling, macronutrient metabolism, negative regulation of gluconeogenesis, regulation of autophagy [27]	N/A
Stard9, star-related lipid transfer domain containing 9	2.14	0.046	Enables ATP hydrolase activity; involved in microtube movement and spindle assembly	668880
Tbc1d10a, tbc1 domain family member 10a	0.91	<0.0001	Enables GTPase activator activity and PDZ domain binding activity; positive regulator of hydrolase activity; involved in protein transport	103724
Tmc8, transmembrane channel like 8	−1.72	<0.0001	Enables TNF-alpha binding activity, zinc homeostasis; regulation of extrinsic apoptosis signaling death receptors	217356
Tpt1-ps3, tumor protein, translationally controlled	4.17	0.005	Cell growth and proliferation	100043703
Uso1 *, uso1 vesicle docking factor	2.10	<0.0001	Regulation of insulin stimulus, secretory granule localization, small GTPase signal transduction, intracellular protein transport, glucose response marker [28]	56041
Vps13b, vacuolar protein sorting 13 homolog b	0.84	<0.0001	Involved in protein transport	666173
Xpo4, exportin 4	1.07	<0.0001	Enables nuclear export signal receptor activity	57258

Data are log_2 fold change. Change in total RNA expression within isolated total mammary epithelial cells of transgenic *fat-1* mice, which genetically reflects lifelong *n*-3 polyunsaturated fatty acid (*n*-3 PUFA) exposure [10], compared to female wildtype (WT) mice on a *n*-6 polyunsaturated fatty acid (*n*-6 PUFA)-rich diet aged 6 to 8 weeks. Three pooled samples from *fat-1* and WT mice underwent RNA sequencing, performed using the Illumina platform (paired-end reads of a 100 bp sequence), and mapped to reference genome GRCM38 using HISAT2 [18]. Differential expression analysis was performed using *DESeq2* followed by multiple testing correction using the Benjamani–Hochberg procedure to control for false discovery rate (FDR) at 5% [21]. Using a threshold of the FDR-adjusted *p*-value of 0.05 revealed 101 genes that were differentially expressed by *DESeq2*. Of these, 45 genes with a known nomenclature are reported in this Table. * Genes that were identified as differentially expressed compared to the WT using the DeSeq2 method after the Benjamini–Hocheburgh adjustment for multiple-hypothesis testing. Gene functions were found using the Gene ID description from the NIH National Library of Medicine on 4 November 2023. References were provided for additional gene functions not listed in the NIH National Library of Medicine.

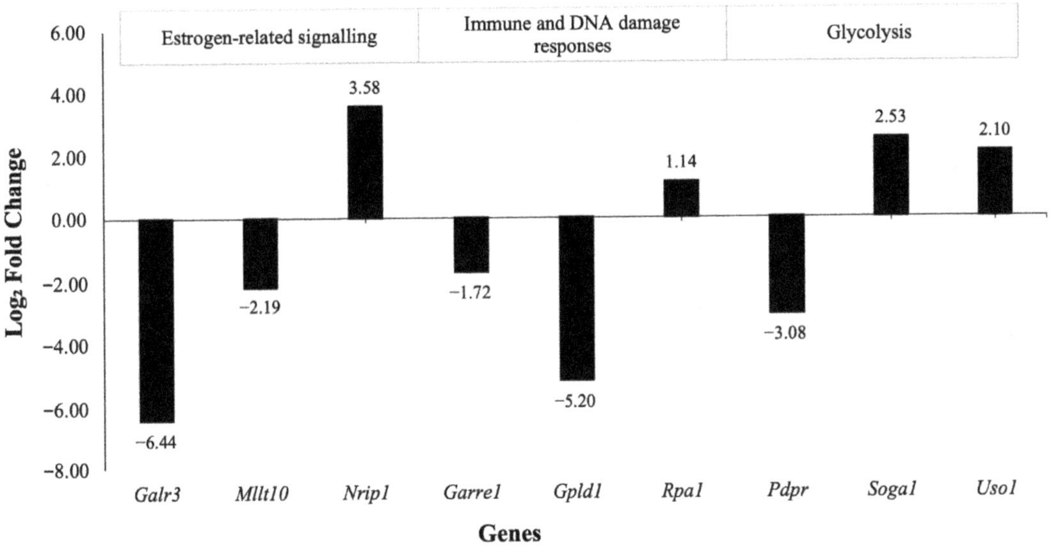

Figure 1. Comparison of concordant and differentially expressed genes ($p \leq 0.05$) across both *edgeR* and *DESeq2* in isolated mammary epithelial cells of 6- to 8-week-old female transgenic *fat-1* mice (n = 3) relative to WT mice (n = 3). Data are *DESeq2*. See methods and details in Table 1.

4. Discussion

This study has identified, in pubertal mammary epithelial cells, using RNA-Seq, differential expression of genes attributable to the presence or absence of *n*-3 PUFA in the transgenic *fat-1* mouse model. Using the transgenic *fat-1* mouse model [9,10], which is capable of de novo *n*-3 PUFA synthesis, this genetic approach provides evidence of within-cell gene expression changes attributable to *n*-3 PUFA. Thus, providing novel insight into how *n*-3 PUFA affects mammary gland development, a key lifecycle stage that potentially can reduce the development of mammary tumors.

At the onset of puberty, estrogen activity largely promotes the rapid growth and expansion of the ducts as TEBs invade to the edge of the fat pad to mature the ductal network [7]. Estrogen binds to its receptor ERα acting on the epithelium to sustain proliferation during ductal elongation [7]. Elevated prepubertal estrogen levels cause earlier puberty onset and increase the risk of BC potentially due to an increased number of proliferative TEBs [8]. Previous studies have demonstrated that prepubertal exposure to *n*-3 PUFA delays puberty onset, lowers estradiol and proliferation and reduces TEBs [9,10]. Consistent with these findings, this study found that *n*-3 PUFA decreases expression of the estrogen-responsive *Galr3*, which has been shown to increase five-fold with estradiol treatment in the female rat anterior pituitary gland [22]. While complete understanding of *Galr3* remains elusive, studies suggest *Galr3* exerts its function through Gi/Go to G proteins, leading to the inhibition of adenyl cyclase that perturbs the phosphorylation of CREB and is more expressed during proliferation of the mammary gland [29,30]. Thus, these changes in *Galr3* could suggest that *n*-3 PUFA contributes to reducing estrogen and proliferation. However, our lab more recently reported that lifelong *n*-3 PUFA exposure also increases ERα protein expression and the relative number of luminal mammary epithelial cells, suggesting additional effects on mammary epithelial cell differentiation [10]. Consistent with this finding, this study found that lifelong *n*-3 PUFA exposure decreases *Mllt10/AF10*, a cofactor of the disruptor silencing 1 like (DOT1L) responsible for di- and tri-histone H3-lysine 79 (H3K79) methylation [24,31]. One study found that H3K79 methylation depletes as mammary epithelial cells lose lineage commitment and become dedifferentiated [32]. While further investigation is required, this effect was hypothesized to occur through a hormonal and/or paracrine mechanism [32], which may be possible as a more recent study has shown that estrogen treatment enhances DOT1L and ERα interaction [33]. Consistent with this hypothesis, this study revealed a significant upregulation of *Nrip1/RIP140*, a critical estrogen-signaling mediator of ductal morphogenesis during pubertal mammary gland development [25]. A previous study reported that *RIP140* functions as a cofactor that is recruited with ERα to promote ERα-targeted gene expression [25]. The loss of *RIP140*, as seen in *RIP140* knockout mice, reduced luminal epithelial cells and impaired TEB formation during puberty [25]. While *RIP140* knockout mice were found to have lower numbers of TEBs than *RIP140* transgenic mice, due to impaired TEB formation, it was shown that overexpression of *RIP140* increases the number of alveolar buds that differentiate from TEBs [25]. Thus, these findings suggest that lifelong *n*-3 PUFA exposure increases estrogen-related signaling pathways under lower estrogen as a potential mechanism during pubertal mammary gland development.

While increasing estrogen-related signaling has also been shown to promote aerobic glycolysis in purified primary mammary mouse epithelial cells [34], limited evidence supports the hypothesis that *n*-3 PUFA increases the estrogen-related signaling that results in the promotion of glycolytic activity. Notably, a study in the triple-negative BC cell line MDA-MB-231 demonstrated that the *n*-3 PUFA, docosahexaenoic acid (DHA), decreased glycolytic activity and mitochondrial respiration [35]. In contrast, the authors also found that applying low concentrations of DHA (15 μM and 25 μM) to the human epithelial cell line of MCF-10A resulted in increased glycolytic utilization [35]. Consistent with these findings, this study shows that *n*-3 PUFA induces a glycolytic profile in healthy mammary epithelial cells during puberty. In mammary epithelial cells, lifelong *n*-3 PUFA exposure upregulated *Soga1*, which promotes glycolysis and reduces gluconeogenesis [27], and *Uso1*, which is highly expressed in glucose response [28], along with the downregulation of *Pdpr*, a gene involved in progressing acetyl-CoA into the citric acid cycle [36]. This glycolytic-like profile may align with evidence suggesting that *n*-3 PUFA can prolong the G1 phase of the cell cycle in embryonic stem cells, as glycolysis can occur during the G1 phase [37]. This halt in cell cycle progression in G1 has been reported to occur despite the presence of oxygen and functional mitochondria via aerobic glycolysis, also known as the Warburg effect [38,39]. More recently, the Warburg effect was found to occur not only in cancer cells but also uniquely in normal proliferating cells, maintaining intracellular pH during cell division [38]. Thus, these findings collectively suggest that *n*-3 PUFA could adaptively modulate metabolic activity in mammary epithelial cells differently in healthy and disease conditions, thus warranting further investigation.

The effects of *n*-3 PUFA mediating the immune and DNA damage response have been widely reported. Previous studies have suggested that *n*-3 PUFA have anticancer effects through the immune system, such as by suppressing CD4+ T-cell activation and reorganizing cell signaling [40]. Consistent with this, we found that lifelong *n*-3 PUFA exposure downregulates *Glpd1*, a phospholipase D1 glycosylphosphatidylinositol anchor, in which splenic CD4+ T cells were found to be downregulated in phospholipase D1 (PLD1) knockout mice [23]. Furthermore, phospholipase D inhibition was shown to reduce BC invasion by lowering PLD1 [23], which is consistent with a study that shows that treatment with essential *n*-6 PUFA increases PLD1 and BC invasion in vitro [41]. However, we also observed an upregulation in *Garre1*, a master regulator of the CCR4-Not complex binding activity with critical roles in the immune system [42–44]. The diverse functions of the CCR4-Not complex, such as cell cycle control, chromatin modification, and transcription activity, allow for the rapid adaption of gene expression in response to environmental changes [42]. A previous study showed that the CCR4-Not complex regulates genomic stability [43], making it a target of interest for cancer therapies [44]. In tandem, we also found that *n*-3 PUFA *Rpa1*, which supports genomic integrity with essential roles in DNA replication, recombination, and repair [26], is decreased in HER2-positive BC [45]. *Rpa1* is also highly expressed in the immune system of mice, particularly in the lymph nodes and spleen, where it plays a crucial role in maintaining T-cell homeostasis [46]. Thus, these findings suggest that *n*-3 PUFA could have critical effects on mammary epithelial cells that are important for maintaining immune function during development, warranting further investigation.

This pilot study has strengths and limitations that should be considered. These results are exclusive to total mammary epithelial cells in 6- to 8-week-old female *fat-1* mice, which consist of both luminal and myoepithelial cells. This study also had a relatively small sample size of three per group. Nevertheless, we used two robust bioinformatic techniques appropriate for small sample sizes [47]. Additionally, the oestrous stage of each mouse was not determined, which could add variability to the results due to higher progesterone and cell proliferation levels during diestrus [48]. Lastly, this study utilized total RNA, including both mRNA and non-coding RNA; however, total RNA has the largest gene library available for clinical use [49,50]. Despite these limitations, this study demonstrated the utility of determining cell-specific changes in gene expression by RNA-Seq. Further, the use of the *fat-1* model made it possible to study the effects of *n*-3 PUFA on mammary gland development without the potential confounding effects of dietary intake [9,10].

5. Conclusions

In conclusion, this study advances our fundamental knowledge of the role of *n*-3 PUFA in pubertal mammary gland development in female *fat-1* mice utilizing cutting edge RNA-Seq technology and bioinformatics approaches. These findings suggest that lifelong *n*-3 PUFA exposure may have long-term protective effects for BC prevention mediated at the level of mammary epithelial cells. Therefore, future studies are warranted to investigate the effects of *n*-3 PUFA on mammary gland development within specific cell types, leading to a better understanding of how diet contributes to reduced BC risk later in life.

Supplementary Materials: The following supporting information can be downloaded at: https://www.mdpi.com/article/10.3390/nu16223925/s1.

Author Contributions: C.D.C.B.: conceptualization, writing—original draft, data curation, writing—review and editing; R.A.: data curation, visualization, conceptualization, writing—original draft, writing—review and editing; L.M.H.: investigation, writing—review and editing; W.T.: methodology, writing—review and editing; J.X.K.: resources, writing—review and editing; S.S.: formal analysis, methodology, resources, writing—review and editing; D.W.L.M.: project administration, conceptualization, methodology, resources, writing—review and editing, funding acquisition. All authors have read and agreed to the published version of the manuscript.

Funding: Funding from the Natural Sciences and Engineering Research Council of Canada [grant # RGPIN-2022-04027] provided to D.W.L. Ma and S. Subedi from Canada Research Chair Program to S. Subedi. C. D.C. Buchanan and R. Ashraf are funded by Ontario Graduate Scholarship (OGS).

Institutional Review Board Statement: This study was approved by the Animal Care Committee of the University of Guelph under the governance of the Canadian Council on Animal Care (Approval Code: Animal Utilization Protocol 4417, Approval Date: July 2020).

Data Availability Statement: The authors declare that all data supporting the study findings are within the article, and the RNA sequencing data have been deposited in the Gene Expression Omnibus (GEO) database under accession code GSE281867 (Available online: https://www.ncbi.nlm.nih.gov/geo/query/acc.cgi?acc=GSE281867 (accessed on 14 November 2024).

Conflicts of Interest: The authors declare no conflicts of interest.

References

1. Bray, F.; Laversanne, M.; Sung, H.; Ferlay, J.; Siegel, R.L.; Soerjomataram, I.; Jemal, A. Global Cancer Statistics 2022: GLOBOCAN Estimates of Incidence and Mortality Worldwide for 36 Cancers in 185 Countries. *CA Cancer J. Clin.* **2024**, *74*, 229–263. [CrossRef] [PubMed]
2. Kerschbaum, E.; Nüssler, V. Cancer Prevention with Nutrition and Lifestyle. *Visc. Med.* **2019**, *35*, 204–209. [CrossRef] [PubMed]
3. Haraldsdottir, A.; Steingrimsdottir, L.; Valdimarsdottir, U.A.; Aspelund, T.; Tryggvadottir, L.; Harris, T.B.; Launer, L.J.; Mucci, L.A.; Giovannucci, E.L.; Adami, H.-O.; et al. Early Life Residence, Fish Consumption, and Risk of Breast Cancer. *Cancer Epidemiol. Biomark. Prev.* **2017**, *26*, 346–354. [CrossRef] [PubMed]

4. MacLennan, M.B.; Clarke, S.E.; Perez, K.; Wood, G.A.; Muller, W.J.; Kang, J.X.; Ma, D.W.L. Mammary Tumor Development Is Directly Inhibited by Lifelong N-3 Polyunsaturated Fatty Acids. *J. Nutr. Biochem.* **2013**, *24*, 388–395. [CrossRef] [PubMed]
5. Zou, Z.; Bidu, C.; Bellenger, S.; Narce, M.; Bellenger, J. N-3 Polyunsaturated Fatty Acids and HER2-Positive Breast Cancer: Interest of the Fat-1 Transgenic Mouse Model over Conventional Dietary Supplementation. *Biochimie* **2014**, *96*, 22–27. [CrossRef]
6. Plante, I.; Stewart, M.K.G.; Laird, D.W. Evaluation of Mammary Gland Development and Function in Mouse Models. *J. Vis. Exp.* **2011**, *53*, 2828. [CrossRef]
7. Paine, I.S.; Lewis, M.T. The Terminal End Bud: The Little Engine That Could. *J. Mammary Gland. Biol. Neoplasia* **2017**, *22*, 93–108. [CrossRef]
8. Krisanits, B.; Randise, J.F.; Burton, C.E.; Findlay, V.J.; Turner, D.P. Pubertal Mammary Development as a "Susceptibility Window" for Breast Cancer Disparity. *Adv. Cancer Res.* **2020**, *146*, 57–82. [CrossRef]
9. Anderson, B.M.; MacLennan, M.B.; Hillyer, L.M.; Ma, D.W.L. Lifelong Exposure to N-3 PUFA Affects Pubertal Mammary Gland Development. *Appl. Physiol. Nutr. Metab.* **2014**, *39*, 699–706. [CrossRef]
10. Hillyer, L.M.; Kang, J.X.; Ma, D.W.L. Lifelong N-3 Polyunsaturated Fatty Acid Exposure Modulates Size of Mammary Epithelial Cell Populations and Expression of Caveolae Resident Proteins in Fat-1 Mice. *Nutrients* **2019**, *11*, 2477. [CrossRef]
11. Wu, J.; Luo, J.; He, Q.; Xia, Y.; Tian, H.; Zhu, L.; Li, C.; Loor, J.J. Docosahexaenoic Acid Alters Lipid Metabolism Processes via H3K9ac Epigenetic Modification in Dairy Goat. *J. Agric. Food Chem.* **2023**, *71*, 8527–8539. [CrossRef] [PubMed]
12. Pal, B.; Chen, Y.; Milevskiy, M.J.G.; Vaillant, F.; Prokopuk, L.; Dawson, C.A.; Capaldo, B.D.; Song, X.; Jackling, F.; Timpson, P.; et al. Single Cell Transcriptome Atlas of Mouse Mammary Epithelial Cells across Development. *Breast Cancer Res.* **2021**, *23*, 69. [CrossRef] [PubMed]
13. Ma, D.W.L.; Seo, J.; Davidson, L.A.; Callaway, E.S.; Fan, Y.-Y.; Lupton, J.R.; Chapkin, R.S. N-3 PUFA Alter Caveolae Lipid Composition and Resident Protein Localization in Mouse Colon. *FASEB J.* **2004**, *18*, 1040–1042. [CrossRef] [PubMed]
14. Kang, J.X. Fat-1 Transgenic Mice: A New Model for Omega-3 Research. *Prostaglandins Leukot. Essent. Fat. Acids* **2007**, *77*, 263–267. [CrossRef] [PubMed]
15. Prater, M.; Shehata, M.; Watson, C.J.; Stingl, J. Enzymatic Dissociation, Flow Cytometric Analysis, and Culture of Normal Mouse Mammary Tissue. *Methods Mol. Biol.* **2013**, *946*, 395–409. [CrossRef]
16. Bettedi, L.; Yan, A.; Schuster, E.; Alic, N.; Foukas, L.C. Increased Mitochondrial and Lipid Metabolism Is a Conserved Effect of Insulin/PI3K Pathway Downregulation in Adipose Tissue. *Sci. Rep.* **2020**, *10*, 3418. [CrossRef]
17. Pertea, M.; Pertea, G.M.; Antonescu, C.M.; Chang, T.-C.; Mendell, J.T.; Salzberg, S.L. StringTie Enables Improved Reconstruction of a Transcriptome from RNA-Seq Reads. *Nat. Biotechnol.* **2015**, *33*, 290–295. [CrossRef]
18. Kim, D.; Langmead, B.; Salzberg, S.L. HISAT: A Fast Spliced Aligner with Low Memory Requirements. *Nat. Methods* **2015**, *12*, 357–360. [CrossRef]
19. Robinson, M.D.; McCarthy, D.J.; Smyth, G.K. edgeR: A Bioconductor Package for Differential Expression Analysis of Digital Gene Expression Data. *Bioinformatics* **2010**, *26*, 139–140. [CrossRef]
20. Love, M.I.; Huber, W.; Anders, S. Moderated Estimation of Fold Change and Dispersion for RNA-Seq Data with DESeq2. *Genome Biol.* **2014**, *15*, 550. [CrossRef]
21. Benjamini, Y.; Hochberg, Y. Controlling the False Discovery Rate: A Practical and Powerful Approach to Multiple Testing. *J. R. Stat. Soc. Ser. B* **1995**, *57*, 289–300. [CrossRef]
22. Gajewska, A.; Zielinska-Gorska, M.; Wasilewska-Dziubinska, E.; Baran, M.; Kotarba, G.; Gorski, K. Pituitary Galaninergic System Activity in Female Rats: The Regulatory Role of Gonadal Steroids. *J. Physiol. Pharmacol.* **2016**, *67*, 423–429. [PubMed]
23. Zhu, M.; Foreman, D.; O'Brien, S.A.; Jin, Y.; Zhang, W. Phospholipase D in T Cell Receptor-Mediated Signaling and T Cell Activation. *J. Immunol.* **2018**, *200*, 2165–2173. [CrossRef] [PubMed]
24. Uğurlu-Çimen, D.; Odluyurt, D.; Sevinç, K.; Özkan-Küçük, N.E.; Özçimen, B.; Demirtaş, D.; Enüstün, E.; Aztekin, C.; Philpott, M.; Oppermann, U.; et al. AF10 (MLLT10) Prevents Somatic Cell Reprogramming through Regulation of DOT1L-Mediated H3K79 Methylation. *Epigenet. Chromatin* **2021**, *14*, 32. [CrossRef] [PubMed]
25. Nautiyal, J.; Steel, J.H.; Mane, M.R.; Oduwole, O.; Poliandri, A.; Alexi, X.; Wood, N.; Poutanen, M.; Zwart, W.; Stingl, J.; et al. The Transcriptional Co-Factor RIP140 Regulates Mammary Gland Development by Promoting the Generation of Key Mitogenic Signals. *Development* **2013**, *140*, 1079–1089. [CrossRef]
26. Dueva, R.; Iliakis, G. Replication Protein A: A Multifunctional Protein with Roles in DNA Replication, Repair and Beyond. *NAR Cancer* **2020**, *2*, zcaa022. [CrossRef]
27. Kruse, R.; Krantz, J.; Barker, N.; Coletta, R.L.; Rafikov, R.; Luo, M.; Højlund, K.; Mandarino, L.J.; Langlais, P.R. Characterization of the CLASP2 Protein Interaction Network Identifies SOGA1 as a Microtubule-Associated Protein. *Mol. Cell. Proteom.* **2017**, *16*, 1718–1735. [CrossRef]
28. Martens, G.A.; Jiang, L.; Verhaeghen, K.; Connolly, J.B.; Geromanos, S.G.; Stangé, G.; Van Oudenhove, L.; Devreese, B.; Hellemans, K.H.; Ling, Z.; et al. Protein Markers for Insulin-Producing Beta Cells with Higher Glucose Sensitivity. *PLoS ONE* **2010**, *5*, e14214. [CrossRef]
29. Gopalakrishnan, L.; Chatterjee, O.; Raj, C.; Pullimamidi, D.; Advani, J.; Mahadevan, A.; Keshava Prasad, T.S. An Assembly of Galanin–Galanin Receptor Signaling Network. *J. Cell Commun. Signal.* **2021**, *15*, 269–275. [CrossRef]
30. Naylor, M.J.; Ginsburg, E.; Iismaa, T.P.; Vonderhaar, B.K.; Wynick, D.; Ormandy, C.J. The Neuropeptide Galanin Augments Lobuloalveolar Development. *J. Biol. Chem.* **2003**, *278*, 29145–29152. [CrossRef]

31. Wille, C.K.; Neumann, E.N.; Deshpande, A.J.; Sridharan, R. DOT1L Interaction Partner AF10 Controls Patterning of H3K79 Methylation and RNA Polymerase II to Maintain Cell Identity. *Stem Cell Rep.* **2023**, *18*, 2451–2463. [CrossRef] [PubMed]
32. Breindel, J.L.; Skibinski, A.; Sedic, M.; Wronski-Campos, A.; Zhou, W.; Keller, P.J.; Mills, J.; Bradner, J.; Onder, T.; Kuperwasser, C. Epigenetic Reprogramming of Lineage-Committed Human Mammary Epithelial Cells Requires DNMT3A and Loss of DOT1L. *Stem Cell Rep.* **2017**, *9*, 943–955. [CrossRef] [PubMed]
33. Nassa, G.; Salvati, A.; Tarallo, R.; Gigantino, V.; Alexandrova, E.; Memoli, D.; Sellitto, A.; Rizzo, F.; Malanga, D.; Mirante, T.; et al. Inhibition of Histone Methyltransferase DOT1L Silences ERα Gene and Blocks Proliferation of Antiestrogen-Resistant Breast Cancer Cells. *Sci. Adv.* **2019**, *5*, eaav5590. [CrossRef] [PubMed]
34. Lacouture, A.; Jobin, C.; Weidmann, C.; Berthiaume, L.; Bastien, D.; Laverdière, I.; Pelletier, M.; Audet-Walsh, É. A FACS-Free Purification Method to Study Estrogen Signaling, Organoid Formation, and Metabolic Reprogramming in Mammary Epithelial Cells. *Front. Endocrinol.* **2021**, *12*, 672466. [CrossRef] [PubMed]
35. Mouradian, M.; Kikawa, K.D.; Dranka, B.P.; Komas, S.M.; Kalyanaraman, B.; Pardini, R.S. Docosahexaenoic Acid Attenuates Breast Cancer Cell Metabolism and the Warburg Phenotype by Targeting Bioenergetic Function. *Mol. Carcinog.* **2015**, *54*, 810–820. [CrossRef]
36. Rajagopalan, K.N.; Egnatchik, R.A.; Calvaruso, M.A.; Wasti, A.T.; Padanad, M.S.; Boroughs, L.K.; Ko, B.; Hensley, C.T.; Acar, M.; Hu, Z.; et al. Metabolic Plasticity Maintains Proliferation in Pyruvate Dehydrogenase Deficient Cells. *Cancer Metab.* **2015**, *3*, 7. [CrossRef]
37. Wei, Z.; Li, D.; Zhu, L.; Yang, L.; Chen, C.; Bai, C.; Li, G. Omega 3 Polyunsaturated Fatty Acids Inhibit Cell Proliferation by Regulating Cell Cycle in Fad3b Transgenic Mouse Embryonic Stem Cells. *Lipids Health Dis.* **2018**, *17*, 210. [CrossRef]
38. Sun, H.; Chen, L.; Cao, S.; Liang, Y.; Xu, Y. Warburg Effects in Cancer and Normal Proliferating Cells: Two Tales of the Same Name. *Genom. Proteom. Bioinform.* **2019**, *17*, 273–286. [CrossRef]
39. Vander Heiden, M.G.; Cantley, L.C.; Thompson, C.B. Understanding the Warburg Effect: The Metabolic Requirements of Cell Proliferation. *Science* **2009**, *324*, 1029–1033. [CrossRef]
40. Hou, T.Y.; McMurray, D.N.; Chapkin, R.S. Omega-3 Fatty Acids, Lipid Rafts, and T Cell Signaling. *Eur. J. Pharmacol.* **2016**, *785*, 2–9. [CrossRef]
41. Diaz-Aragon, R., Ramirez-Ricardo, J.; Cortes-Reynosa, P.; Simoni-Nieves, A.; Gomez Quiroz, L. E.; Perez Salazar, E. Role of Phospholipase D in Migration and Invasion Induced by Linoleic Acid in Breast Cancer Cells. *Mol. Cell. Biochem.* **2019**, *457*, 119–132. [CrossRef] [PubMed]
42. Miller, J.E.; Reese, J.C. Ccr4-Not Complex: The Control Freak of Eukaryotic Cells. *Crit. Rev. Biochem. Mol. Biol.* **2012**, *47*, 315–333. [CrossRef] [PubMed]
43. Hagkarim, N.C.; Hajkarim, M.C.; Suzuki, T.; Fujiwara, T.; Winkler, G.S.; Stewart, G.S.; Grand, R.J. Disruption of the Mammalian Ccr4–Not Complex Contributes to Transcription-Mediated Genome Instability. *Cells* **2023**, *12*, 1868. [CrossRef] [PubMed]
44. Robles, O.; Jackson, J.J.; Marshall, L.; Talay, O.; Chian, D.; Cutler, G.; Diokno, R.; Hu, D.X.; Jacobson, S.; Karbarz, E.; et al. Novel Piperidinyl-Azetidines as Potent and Selective CCR4 Antagonists Elicit Antitumor Response as a Single Agent and in Combination with Checkpoint Inhibitors. *J. Med. Chem.* **2020**, *63*, 8584–8607. [CrossRef] [PubMed]
45. Park, S.; Lee, K.; Ju, J.; Kim, J.; Noh, D.-Y.; Lee, T.; Shin, I. Protein Expression Profiling of Primary Mammary Epithelial Cells Derived from MMTV-Neu Mice Revealed That HER2/NEU-Driven Changes in Protein Expression Are Functionally Clustered. *IUBMB Life* **2010**, *62*, 41–50. [CrossRef]
46. Song, J.; Zhang, X.; Yin, Y.; Guo, M.; Zhao, X.; Wang, L.; Ren, C.; Yin, Y.; Zhang, X.; Deng, X.; et al. Loss of RPA1 Impairs Peripheral T Cell Homeostasis and Exacerbates Inflammatory Damage through Triggering T Cell Necroptosis. *Adv. Sci.* **2023**, *10*, e2206344. [CrossRef]
47. Khang, T.F.; Lau, C.Y. Getting the Most out of RNA-Seq Data Analysis. *PeerJ* **2015**, *3*, e1360. [CrossRef]
48. Shehata, M.; Waterhouse, P.D.; Casey, A.E.; Fang, H.; Hazelwood, L.; Khokha, R. Proliferative Heterogeneity of Murine Epithelial Cells in the Adult Mammary Gland. *Commun. Biol.* **2018**, *1*, 111. [CrossRef]
49. Kukurba, K.R.; Montgomery, S.B. RNA Sequencing and Analysis. *Cold Spring Harb. Protoc.* **2015**, *2015*, 951–969. [CrossRef]
50. Guo, Y.; Zhao, S.; Sheng, Q.; Guo, M.; Lehmann, B.; Pietenpol, J.; Samuels, D.C.; Shyr, Y. RNAseq by Total RNA Library Identifies Additional RNAs Compared to Poly(A) RNA Library. *BioMed Res. Int.* **2015**, *2015*, 862130. [CrossRef]

nutrients

MDPI

Article

The Impact of Haplotypes of the *FTO* Gene, Lifestyle, and Dietary Patterns on BMI and Metabolic Syndrome in Polish Young Adult Men

Sylwia Górczyńska-Kosiorz [1,*], Mateusz Lejawa [2], Marcin Goławski [2], Agnieszka Tomaszewska [3], Martyna Fronczek [2], Beata Maksym [2], Maciej Banach [4,5] and Tadeusz Osadnik [2]

1 Department of Internal Medicine, Diabetology and Nephrology, Faculty of Medical Sciences in Zabrze, Medical University of Silesia, 40-055 Katowice, Poland
2 Department of Pharmacology, Faculty of Medical Sciences in Zabrze, Medical University of Silesia, 40-055 Katowice, Poland; mateusz.lejawa@sum.edu.pl (M.L.); d201182@365.sum.edu.pl (M.G.); mfronczek@sum.edu.pl (M.F.); bmaksym@sum.edu.pl (B.M.); tadeusz.osadnik@sum.edu.pl (T.O.)
3 Prenatal Diagnostic and Genetic Clinic, Medical Center, Medical University of Silesia, 41-800 Zabrze, Poland; atomaszewska@szpital.zabrze.pl
4 Department of Preventive Cardiology and Lipidology, Medical University of Lodz (MUL), 90-549 Lodz, Poland; maciej.banach@icloud.com
5 Ciccarone Center for the Prevention of Cardiovascular Disease, Division of Cardiology, Department of Medicine, Johns Hopkins University School of Medicine, 600 N. Wolfe St., Carnegie 591, Baltimore, MD 21287, USA
* Correspondence: skosiorz@sum.edu.pl

Abstract: Background: Variants in fat mass and the obesity-associated protein (*FTO*) gene have long been recognized as the most significant genetic predictors of body fat mass and obesity. Nevertheless, despite the overall evidence, there are conflicting reports regarding the correlation between different polymorphisms of the *FTO* gene and body mass index (BMI). Additionally, it is unclear whether *FTO* influences metabolic syndrome (MetS) through mechanisms other than BMI's impact. In this work, we aimed to analyze the impact of the following *FTO* polymorphisms on the BMI as well as MetS components in a population of young adult men. Methods: The patient group consisted of 279 Polish young adult men aged 28.92 (4.28) recruited for the MAGNETIC trial. The single-nucleotide polymorphisms (SNPs), located in the first intron of the *FTO* gene, were genotyped, and the results were used to identify "protective" and "risk" haplotypes and diplotypes based on the literature data. Laboratory, as well as anthropometric measurements regarding MetS, were performed. Measured MetS components included those used in the definition in accordance with the current guidelines. Data regarding dietary patterns were also collected, and principal components of the dietary patterns were identified. Results: No statistically significant correlations were identified between the analyzed *FTO* diplotypes and BMI ($p = 0.53$) or other MetS components (waist circumference $p = 0.55$; triglycerides $p = 0.72$; HDL cholesterol $p = 0.33$; blood glucose $p = 0.20$; systolic blood pressure $p = 0.06$; diastolic blood pressure $p = 0.21$). Stratification by the level of physical activity or adherence to the dietary patterns also did not result in any statistically significant result. Conclusions: Some studies have shown that *FTO* SNPs such as rs1421085, rs1121980, rs8050136, rs9939609, and rs9930506 have an impact on the BMI or other MetS components; nevertheless, this was not replicated in this study of Polish young adult males.

Keywords: *FTO*; haplotypes; gene polymorphism; genetic factors; BMI; metabolic syndrome; diet quality; lifestyle

1. Introduction

Obesity is defined as an excess adipose tissue accumulation. This is often defined as a body mass index (BMI) greater or equal to 30 for adults. In 2016, over 650 million people were obese, and the prevalence of obesity has tripled since 1975. Obesity increases the risk of cardiovascular diseases such as coronary artery disease and stroke, diabetes mellitus (DM), musculoskeletal disorders such as osteoarthritis, and some cancers, including endometrial, breast, ovarian, and prostate cancer [1,2].

Obesity is a component of metabolic syndrome (MetS), an interconnected group of risk factors influencing the development of atherosclerotic cardiovascular disease. Numerous definitions exist [3,4], but one of them was formulated in the Joint Interim Statement of the International Diabetes Federation Task Force on Epidemiology and Prevention; the National Heart, Lung, and Blood Institute; the American Heart Association; the World Heart Federation; the International Atherosclerosis Society; and the International Association for the Study of Obesity. For an adult to be diagnosed with MetS according to this definition, they have to fulfill at least three of the five criteria of abdominal obesity, hypertriglyceridemia, low HDL cholesterol, hypertension, and impaired glucose metabolism [5]. Although closely connected with obesity, MetS was associated with an increased mortality, even in normal-weight individuals [6].

The etiology of obesity is complex and combines genetic, behavioral, environmental, psychological, social, and even cultural influences [7]. Twin, family, and adoption studies suggest that as much as 40% to 70% of the body weight index variation is explained by genetic factors [8,9]. Genetic influences on obesity may be divided into two distinct categories. The first one is monogenic obesity, which is typically early-onset and severe. It is caused by a mutation in a single gene, such as one encoding leptin, leptin receptor, or melanocortin receptor 4 [10,11]. The second, more frequent type is polygenic obesity, caused by multiple mutations or common population variants in several genes. It increases individual susceptibility to environmental factors that cause obesity and is also characterized by later onset and reduced severity [10].

Although many genes are known to influence polygenic obesity, polymorphisms of fat mass and the obesity-associated protein (*FTO*) gene were quickly recognized as highly influential regarding BMI and body adiposity [12,13]. FTO is a Fe (II)- and 2-oxoglutarate-dependent N6-methyladenosine demethylase, which removes methyl groups from N6-methyladenosine in mRNA and thusly reverses a common eukaryotic post-translational modification motif [14]. Single-nucleotide polymorphisms (SNPs) in *FTO* gene intron 1 have been shown to affect *FTO* expression [15,16], while FTO protein itself appears to influence adipogenesis by increasing the abundance of a short splice variant of RUNX1 translocation partner 1 (RUNX1T1) [17–20]. FTO controls the splicing of the adipogenic regulatory factor RUNX1T1 by regulating m6a. M6a affects several cellular processes, such as gene regulation and degradation, translation, or transport, and modulates alternative splicing and mRNA stability. An inverse correlation has been demonstrated between FTO and m6a, which is implicated in adipogenesis and demonstrates how FTO directly modulates obesity at the level of m6A [21]. Interestingly, SNPs of intron 1 of *FTO* have also been suggested to affect the expression of adjacent genes such as Iroquois homeobox 3 (IRX3), Iroquois homeobox 5 (IRX5), and RPGR-Interacting Protein 1-Like (RPGRIP1L) [22–24]. Another possible FTO role in obesity is the regulation of macronutrient intake due to the involvement of *FTO* expressed in the hypothalamus [25–27]. Although the overall impact of *FTO* gene polymorphisms on BMI and body composition was established, the impact and role of specific SNPs are less clear [12,28]. The occurrence of different polymorphic variants (SNPs) in the *FTO* gene may explain why unbalanced results in obesity have been obtained to date. Some reports suggest that specific haplotypes of the *FTO* gene influence rather than just particular SNPs complicating the issue further [21,29].

Thus, in this study, we aimed to determine the influence of polymorphisms in the *FTO* gene on the BMI and MetS components and assess the impact of genetic variants on the type of diet and lifestyles in a homogenous population of young Polish males [30].

2. Materials and Methods

2.1. Study Group

This study is an extension of the MAGNETIC ("Metabolic and Genetic Profiling of Young Adults with and without a Family History of Premature Coronary Heart Disease") project. The study group consisted of randomly selected male participants recruited within the framework of the MAGNETIC project [31]. Briefly, the MAGNETIC project was a case-control study involving young (18–35 years) adults with (cases) and without (controls) a history of premature coronary artery disease (myocardial infarction, coronary artery bypass grafting, or percutaneous coronary intervention before the age of 55 in men and 65 in women) in first-degree relatives (Figure 1).

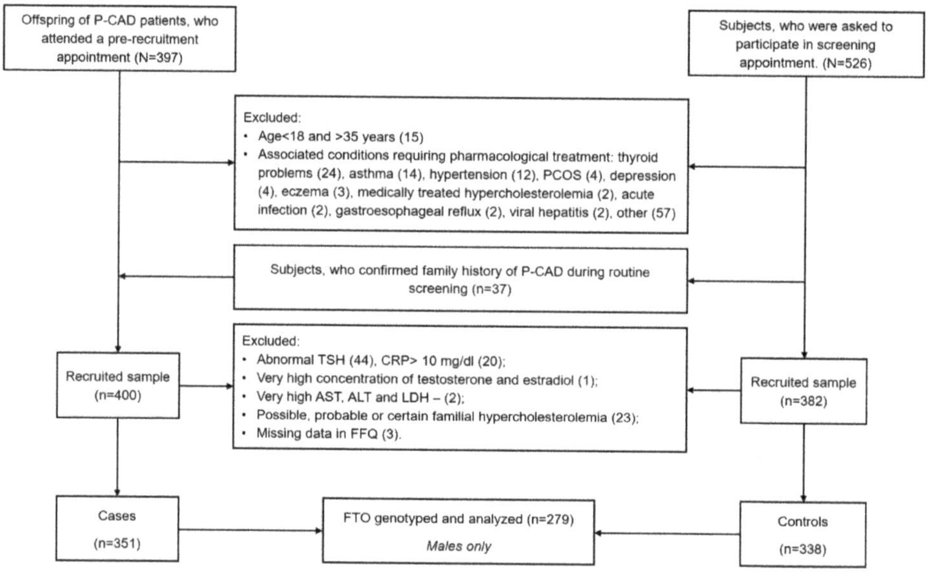

Figure 1. A flowchart of MAGNETIC study.

For the purposes of this study, only men were chosen because there was a higher percentage of obese people with MetS among men than women. None of the study participants was taking lipid-lowering, hypotensive, or anti-diabetic drugs.

The criteria for a MetS diagnosis were based on the Joint Interim Statement Criteria and were [5]:

- Waist circumference ≥80 cm in women and ≥94 cm in men;
- Fasting blood triglycerides > 1.7 mmol/L or treatment of hypertriglyceridemia;
- HDL-C < 1.3 mmol/L in women and <1.0 mmol/L in men or treatment of low HDL-c levels;
- Systolic blood pressure ≥ 130 mmHg or diastolic blood pressure ≥ 85 mm Hg or treatment for hypertension;
- Fasting blood glucose levels ≥ 5.6 mmol/L or treatment of DM.

At least three criteria had to be met for a MetS diagnosis. These criteria were chosen due to being widely used for MetS diagnostics.

2.2. Measurements of Anthropometric Parameters

During an initial visit at the Silesian Center for Heart Disease, trained examiners took measurements of the participants' height, weight, waist, and hip circumference, as well as their systolic and diastolic blood pressure.

2.3. Laboratory Tests

Whole blood samples were collected from each participant 8–10 h after the last meal. Biochemical and immunoenzymatic tests were carried out in a hospital medical diagnostic laboratory. Patient serum measures were performed using a Cobas 6000 analyzer (Roche, Basel, Switzerland). Fibrinogen concentrations were determined using a BCS XP analyzer (Siemens Healthcare, Erlangen, Germany).

2.4. Diet Quality and Lifestyle

To collect an interview on eating habits and nutrition knowledge, the study participants completed two validated questionnaires: the FFQ-6 and KomPAN. The FFQ-6 covers a wide range of products (62 foodstuffs) that are usually consumed in Poland and was previously validated [32]. The KomPAN questionnaire is based on the frequency of consumption of 24 food items in the last year. The KomPAN questionnaire also includes questions about physical activity, smoking, and smoking status. Physical activity was categorized as low (sitting, screen time, reading, light housework, or walking 1–2 h a week), moderate (walking, cycling, moderate exercise, working at home or other light physical activity performed 2–3 h/week), or high (cycling, running, working at home or other sports activities requiring physical effort over 3 h/week) [33].

2.5. Genetic Analysis

DNA was isolated from the study participant's whole blood samples, which were stored below −80 °C. Isolation procedures were performed according to the manufacturer's instructions, using the automated extraction system MagCore®HF 16 and MagCore®GenomicDNA Whole Blood Kit (RBC Bioscience, New Taipei City, Taiwan). The purity and concentration of DNA were determined using a NanoDrop spectrophotometer (Thermo Scientific, Waltham, MA, USA). DNA samples were diluted to a 15 ng/µL concentration in molecular biology-grade water and stored below −30 °C until analysis.

The study sample was genotyped for five variants of the *FTO* gene: rs1421085, rs1121980, rs8050136, rs9939609, and rs9930506 (Table S1). This was achieved using commercially available pre-designed TaqMan SNP genotyping assays (ThermoFisher, Waltham, MA, USA). Each genotyping reaction was conducted in a volume of 20 µL using FastStart Essential DNA Probes Master (Roche, Basel, Switzerland) on the real-time PCR device LifeCycler 96 (Roche, Basel, Switzerland). To ensure the credibility of the results, the genotyping of 10% of the samples was reattempted, achieving 100% repeatability.

2.6. Statistical Analysis

Analysis was performed using R language in the Rstudio environment [34]. In all calculations, results with a $p < 0.05$ were considered statistically significant.

Data were presented as mean and standard deviation (SD) or as absolute (n) and relative (%) frequencies in the tables. The Kruskal–Wallis rank sum or Chi-squared test was used to evaluate group differences.

All norms for the anthropometric and biochemical parameters were based on the Joint Interim Statement criteria [5].

The fast, exact test proposed by Wigginton, Cutler, and Abecasis [35], implemented in the SNPassoc package (version 2.1-0) [36], was used to determine the genotype's consistency with the Hardy–Weinberg equilibrium.

Dietary patterns (DPs) were obtained by principal component analysis (PCA) with normalized varimax rotation based on the FFQ-6 questionnaire. To create these DPs, the frequency of consuming 26 food groups per day was standardized to have a mean of 0 and a standard deviation of 1. We utilized PCA, selecting components to keep, which were based on their interpretability and eigenvalues (>1), along with identifying a breakpoint through the Scree test. The significance of each questionnaire item's contribution to the identified DPs was assessed through factor loadings exceeding $> |0.30|$. The dietary patterns were named according to the variables with the highest loadings for each pattern.

For each individual, a DP score reflecting their adherence to the DP was computed as the sum of the product of food frequency consumption and the factor loading for the 26 food groups. The subjects were categorized into three groups (bottom, middle, and upper tertile) based on the tertile distribution, representing the lowest, moderate, and highest adherence to the DP, respectively [37].

Furthermore, two diet quality scores, the pro-Healthy-Diet-Index (pHDI) and non-Healthy-Diet-Index (nHDI), were calculated based on the KomPAN questionnaires and previously published methodology [38].

3. Results

3.1. Study Group Characteristics

The studied group consisted of 279 young adult males aged 18–36 (Table 1). The MetS criteria were fulfilled for 21.5% of the patients. The average BMI was above the overweight threshold (>25 kg/m^2); 18.28% of the study participants were obese, and 32.98% were overweight. Excessive waist circumference was noted in 31.54% of the patients. Hypertriglyceridemia was noted in 19.71% of the patients, while low blood HDL-C levels were found in 47.67% of the studied population. High blood pressure was found in 57.71% of the studied group. Elevated fasting glucose was present in 17.92% of the studied persons, but none of the participants was diagnosed with DM. All patients were of Caucasian descent and were residents of Upper Silesia.

Table 1. Characteristics of included patients.

Variable	*N* *	Number of Patients (%) or Value (SD) †
Age	279	28.92 (4.28)
Family History of P-CAD (%)	279	178/279 (64%)
Family History of T2DM (%)		99/279 (35%)
Current smoking (vs. past smoker or non-smoker)	279	72/279 (26%)
Physical activity level	279	
Low		73/279 (26%)
Moderate		108/279 (39%)
High		98/279 (35%)
SBP [mmHg]	270	132 (15)
DBP [mmHg]	270	81 (11)
BMI [kg/m^2]	279	26.14 (4.38)
VAI	279	1.58 (2.65)
WC [m]	279	0.89 (0.08)
WHTR	279	0.50 (0.06)
TC [mmol/L]	279	5.12 (1.11)
HDL-C [mmol/L]	279	1.42 (0.37)
LDL-C [mmol/L]	279	3.26 (0.96)
TG [mmol/L]	279	1.42 (1.51)
Lp(a) [nmol/L]	279	42.49 (65.64)
apoA1 [g/L]	276	1.56 (0.25)
apoB [g/L]	277	1.02 (0.60)
Glucose [mmol/L]	279	5.15 (0.45)
HbA1c [%]	278	5.06 (0.25)
hsCRP [mg/dL]	279	1.33 (1.37)
Uric Acid [μmol/L]	279	349.24 (63.96)
Fibrinogen [mg/dL]	278	264.96 (56.89)

P-CAD—premature coronary artery disease; T2DM—diabetes mellitus type 2; SBP—systolic blood pressure; DBP—diastolic blood pressure; BMI—body mass index; VAI—visceral adiposity index; WC—waist circumference; WHTR—waist/hip ratio; TC—total cholesterol; HDL-C—high-density lipoprotein cholesterol; LDL-C—low-density lipoprotein cholesterol; TG—triglycerides; Lp(a)—lipoprotein(a); apoA1—apolipoprotein A1; apoB—apolipoprotein B; HbA1c—glycated hemoglobin; hsCRP—high-sensitivity C-reactive protein. * Full sample size is 279. Smaller *N* numbers indicate missing data. † Values are presented as mean (SD) or *n*/*N* (%).

3.2. FTO SNP

All SNPs were located in the first intron of the *FTO* gene in a region spanning 29,511 bp. They were located between 4252 bp and 9938 bp distance from each other (Table S1). All genotypes of the five analyzed SNPs (rs1421085, rs1121980, rs8050136, rs9939609, and rs9930506) were in Hardy–Weinberg equilibrium. The allele frequencies did not differ significantly based on the BMI category, waist/hip ratio (WHR), HbA1c presence of MetS components, or a family history of T1D or T2D. The genotype frequencies of all tested polymorphisms are presented in Table S2.

During further analysis, seven haplotypes of the analyzed SNPs were identified. The respective alleles in each haplotype are arranged in the following order of SNPs: rs1421085, rs1121980, rs8050136, rs9939609, and rs9930506. The two most common alleles that were found in 95% of the participants were TGCTA (all protective alleles) and CAAAG (all risk alleles) (Table S3). Next, 12 diplotypes were identified (Table S4). In 89% of the participants, the diplotypes consisted of just the two most common haplotypes. According to the nomenclature proposed by Kolackov et al., a diplotype consisting of a pair of TGCTA haplotypes was termed a protective diplotype. The Diplotype consisting of a pair of CAAAG haplotypes was named risk diplotype 2, while the diplotype consisting of one TGCTA and one CAAAG haplotype was named risk diplotype 1 [29]. The protective diplotype was identified in 65 participants, risk diplotype 1 was found in 128 participants, while risk diplotype 2 was identified in 55 participants. Next, the levels of each MetS component in the participants of every identified diplotype were compared (Tables 2 and S6). No significant differences were detected ($p > 0.05$).

Table 2. Analysis of components of metabolic syndrome in relation to *FTO* diplotype (three most common haplotype pairs).

Variable	Protective Diplotype $N = 65$ [1]	"Risk" Diplotype 1 $N = 129$ [1]	"Risk" Diplotype 2 $N = 55$ [1]	*p*-Value [2]
BMI [kg/m^2]	25.86 (4.59)	26.46 (4.43)	26.07 (4.39)	0.53
WC [m]	0.89 (0.12)	0.90 (0.11)	0.90 (0.12)	0.55
TG [mmol/L]	1.26 (0.79)	1.39 (1.42)	1.45 (1.16)	0.72
HDL-C [mmol/L]	1.49 (0.40)	1.37 (0.33)	1.43 (0.39)	0.33
Glucose [mmol/L]	5.08 (0.34)	5.16 (0.47)	5.21 (0.50)	0.20
SBP [mmHg]	131.25 (12.41)	130.81 (14.86)	136.15 (15.33)	0.060
DBP [mmHg]	81.19 (9.25)	80.22 (11.33)	82.56 (12.89)	0.21

BMI—body mass index; SBP—systolic blood pressure; DBP—diastolic blood pressure; HDL-C—high-density lipoprotein cholesterol; TG—triglycerides; WC—waist circumference; [1] Mean (SD); [2] Kruskal–Wallis rank sum test; protective diplotype—TGCTA/TGCTA (rs1421085/rs1121980/rs8050136/rs9939609/rs9930506); risk diplotype 1—TGCTA/CAAAG (rs1421085/rs1121980/rs8050136/rs9939609/rs9930506); risk diplotype 2—CAAAG/CAAAG (rs1421085/rs1121980/rs8050136/rs9939609/rs9930506).

3.3. Dietary Patterns, Dietary Quality, and Physical Activity

PCA was used to identify two dietary patterns (DPs), explaining a cumulative 24% of the variation in the analyzed variables (Table S5). The first "Prudent DP" was characterized by a frequent intake of milk, fermented milk drinks, and curd cheese (0.61), whole grain products (0.67), vegetables (0.67), fish (0.60), fruits (0.54), nuts and seeds (0.53), eggs and egg dishes (0.42), white meat (0.45), and legumes (0.45). The second, "Western DP" was characterized by more frequent consumption of sugar (0.51), refined grain products (0.51), processed meats (0.67), potatoes (0.61), sweets and snacks (0.49), animal fats (0.49), other edible fats (0.46), and sweetened beverages and energy drinks (0.45). When the three most common *FTO* diplotypes were compared regarding the DPs, knowledge of nutrition, or physical activity, no significant differences were found (Table 3); only "Risk diplotype 2" was characterized by a lower pHDI but this was not statistically significant ($p > 0.05$).

Table 3. Analysis of dietary patterns and physical activity in relation to *FTO* diplotypes (three most common haplotype pairs).

Variable	*N*	Protective Diplotype, *N* = 65 [1]	Risk Diplotype 1, *N* = 128 [1]	Risk Diplotype 2, *N* = 55 [1]	*p*-Value [2]
"Prudent" dietary pattern	248				0.61
Lowest adherence to DP		23/65 (35%)	44/128 (34%)	19/55 (35%)	
Moderate adherence to DP		20/65 (31%)	41/128 (32%)	23/55 (42%)	
Highest adherence to DP		22/65 (34%)	43/128 (34%)	13/55 (24%)	
"Western" dietary pattern	248				0.30
Lowest adherence to DP		21/65 (32%)	48/128 (38%)	14/55 (25%)	
Moderate adherence to DP		21/65 (32%)	46/128 (36%)	18/55 (33%)	
Highest adherence to DP		23/65 (35%)	34/128 (27%)	23/55 (42%)	
Physical activity	248				0.68
gentle		17/65 (26%)	33/128 (26%)	14/55 (25%)	
moderate		25/65 (38%)	55/128 (43%)	18/55 (33%)	
vigorous		23/65 (35%)	40/128 (31%)	23/55 (42%)	
nHDI (% points)	246	5.57 (2.42)	5.63 (2.18)	5.53 (2.74)	0.72
pHDI (% points)	247	6.87 (2.55)	6.26 (2.08)	7.18 (2.62)	0.087

DP—dietary patterns; pHDI—pro-Healthy-Diet-Index; nHDI—non-Healthy-Diet-Index; [1] Number of patients (%); [2] Kruskal–Wallis rank sum test or Chi-squared test; protective diplotype—TGCTA/TGCTA (rs1421085/rs1121980/rs8050136/rs9939609/rs9930506); risk diplotype 1—TGCTA/CAAAG (rs1421085/rs1121980/rs8050136/rs9939609/rs9930506; risk diplotype 2—CAAAG/CAAAG (rs1421085/rs1121980/rs8050136/rs9939609/rs9930506).

4. Discussion

In the age of the obesity epidemic, it is important to determine the causes of this phenomenon to enable the development of better ways of preventing and treating this disease.

Previously published analyses based on much larger studies and meta-analyses conducted on a broader population indicate that *FTO* genetic variants correlate with obesity. Some previously published analyses of the Polish population show that the influence of SNPs in the first intron of the *FTO* gene on obesity is modulated by age and gender. In our study, we investigated young adult men, while Sobalska-Kwapis et al. showed an association between obesity risk and SNPs in the *FTO* intron 1 in a group of 2747 men aged 45–50 years. Perhaps the lack of significance in our study was influenced by the properly functioning hormonal system in young men, which could compensate for the genetic predisposition to obesity [39,40]. Other studies concerning the Polish population have also assessed the impact of polymorphisms in the *FTO* gene on BMI and body mass. The study conducted by Piwonska et al. included a group of 3369 patients aged 20–74. In this study, the AA genotype of the rs9939609 variant was significantly correlated with body mass, BMI, waist circumference, and hip circumference [41]. Other analyses conducted by Luczynski et al. in a group of 968 Polish children aged 4–18 years indicated that AA homozygotes of the rs9939609 variant were characterized by increased body mass, BMI, waist circumference, arm circumference, and height [42]. A correlation between obesity and the rs9930506 variant was also demonstrated by Wrzosek et al. in a study of 442 Polish males [40]. Another study, which utilized sequencing of the chromosome 16 fragment encoding the *FTO* gene, has shown a strong correlation between rs9930506 polymorphism and obesity and being overweight in males and the rs1421085 variant and obesity in females. In the same study, a regression model, stratified by age and sex, has shown that allele A of rs9939609 was significantly more frequent in obese persons [43].

Many published analyses confirm the association of intronic polymorphisms of the *FTO* gene with obesity and BMI, but it is important to emphasize that there are also studies similar to those presented here that do not confirm the influence of *FTO* variants on the occurrence of obesity in the Polish population. In one such study, the rs9939609 polymorphism was analyzed in 1097 persons aged 30–80, but a correlation with BMI was found only among males [44]. During the 6-year follow–up, there was no significant impact of rs9939609 polymorphisms on body mass increase [45]. In another study, which was conducted by Kolačkov et al., there was no significant correlation between the *FTO* polymorphisms (rs1421085, rs1121980, rs9939609, and rs9930506) and BMI among persons with a BMI < 25, although there was a tendency for an increased hip circumference among women with haplotype TCGA [29]. This suggests that the influence of *FTO* polymorphisms on BMI in the Polish population may not be universal. We can, therefore, conclude that the relationship between *FTO* and obesity may be more complex than we thought. There may be several reasons why the population studied by our team did not show a clear relationship between the genetic variants in the *FTO* gene and BMI. It is possible that in the case of young Polish men, the genetic variants in the *FTO* gene examined by our team do not significantly affect the body weight of the subjects. Additionally, it should be emphasized that the observations from our study may result from a relatively small study group. To confirm this observation, analyses should be carried out on a much larger group of respondents.

The impact of genetic variants located within the first intron of the *FTO* gene on the occurrence and development of MetS, BMI, and obesity was also analyzed in other populations. In a study conducted by Guclu-Geyik et al., two *FTO* gene polymorphisms, rs9939609 and rs1421085, were evaluated in a population of adult Turks consisting of 1967 adult men and women. The mean age of women was 49.2 ± 11.8 years, and the mean age of men in this study was 50.1 ± 12.0 years. It was shown that both SNPs were strongly associated with MetS in men and with obesity in the studied group of women. Additionally, men carrying the rs1421085 C allele showed an increased risk of insulin resistance and a higher BMI, while this association was not as evident in women [46]. At the same time, it was observed that in some native populations, such as the indigenous population of Samoa and American Samoa, *FTO* variants, e.g., rs1421085, rs1121980, rs8050136, rs9939609, rs9939973, rs17817449, rs3751812, and rs7190492 do not have a statistically significant effect on the BMI [47]. In turn, analyses based on the indigenous Xavante population in Brazil showed a significant correlation between the rs9939609 polymorphism and the body obesity index and being overweight, and at the same time, showed no relationship with obesity, BMI, and waist circumference [48]. These observations indicate that it may be unlikely that the studied population and Polish population, in general, have a genetic make-up that completely negates the influence of *FTO* intron 1 polymorphisms on the BMI, but population-dependent genetic factors that modulate the impact of particular SNPs remain possible.

It should also be noted that physical activity is a factor that may modulate the impact of *FTO* polymorphisms on obesity. In a study involving 420 Brazilians aged 7–17, genotype AA of rs9939609 influenced the BMI and waist circumference only in physically inactive adolescents [49]. On the other hand, in a study of 1152 male Swedes, the same polymorphism showed no statistically significant correlation between the A alleles and BMI. It was observed that increased physical activity reduced the impact of allele A of rs9939609 on the BMI [50]. In our study, the population could be considered quite physically active since 35% of the studied patients reported vigorous physical activity, and a further 39% reported moderate physical activity; it is possible that this partially masks the influence of *FTO* variants on the BMI.

Additionally, in some populations, the impact of *FTO* polymorphisms on the BMI could be masked by dietary habits. In a study conducted by Harbron et al., a relationship of *FTO* polymorphisms (rs1421085 and rs17817449) with physical activity, dietary habits, and BMI was studied in Caucasian patients aged 25–40 with overweight or obesity. There was no direct correlation between the *FTO* polymorphisms and BMI, but two C alleles in rs1421085 increased the feelings of hunger and reduced dietary restraint and control [51]. A study enrolling over 6000 Koreans has shown that rs1121980 A allele carriers with a BMI ≥ 25 consumed more blue fish, organ meat, coffee, and coffee creamer. No such correlation was found in normal-weight males, although overweight and obese female rs1121980 A allele carriers consumed more sweets [52]. Although in our study *FTO* polymorphisms did not appear to affect dietary habits, it is nevertheless possible that the influence of *FTO* polymorphisms would be significant under a different diet than the one our patients had.

Another factor influencing the relationship between *FTO* gene polymorphisms and body mass is age. No correlation between the rs9939609 polymorphisms and BMI was seen in a study of 70-year-old individuals [53]. In another study, there was no impact of several *FTO* polymorphisms (rs9939609, rs8050136, rs1558902, and rs3751812) on the BMI in a group of 401 Han Chinese aged 14–18 [54]. In our study, the population was relatively young, with a mean age of 28.92. This may have affected the findings, although the patients were not elderly or underage, like in some studies. Moreover, some studies that included adolescents showed a correlation between *FTO* polymorphisms and BMI. For example, Kalantari et al. have reported that the GGC haplotype in rs9930506, rs9930501, and rs9932754 is associated with the BMI and related obesity rates in a group of 237 adolescent males aged 12–16 years [55].

In this study, there was also no significant correlation between the *FTO* polymorphisms and components of MetS other than obesity. This is consistent with the hypothesis proposed by other researchers that *FTO* influences lipid, glucose, and blood pressure levels indirectly, chiefly through its influence on body adiposity, body mass, and BMI. This is supported by a study conducted by Freathy et al., who analyzed data from 17,037 Caucasian patients. Upon adjusting for the BMI, there was no correlation between the rs9939609 alleles and metabolic syndrome components [56]. In turn, Doney et al., in a study of 4897 Scottish patients, found that the rs9939609 variants correlated with triglyceride and HDL-C levels, even after adjusting for obesity-related phenotypes [57].

The results presented by our research team in this publication have several strengths and weaknesses. The primary strengths of this work include the enrollment of patients from an understudied Polish population of young adult males and the availability of anthropometric parameters and laboratory test results, as well as physical activity and food intake data. The population was uniform regarding its ethnicity, and it must also be stated that the Polish population is uniform regarding its genetics [30]. Additionally, the study group was homogeneous in terms of age, with a mean age of 28.92 ± 4.28 of the examined males. Nevertheless, this work has some limitations. The analyzed population was rather small, so weak genetic influences could have been statistically insignificant, even if present. Moreover, the food intake and physical activity levels were self-reported, which may introduce some errors, even for validated questionnaires, although there is no ultimate gold standard for food questionnaires tailored to the Polish population. Additionally, the population likely had a genetically increased risk of premature coronary artery disease compared to the general population, and it makes it possible that some alleles of the SNPs of interest were enriched compared to the general population.

5. Conclusions

In this work, there is no statistically significant correlation between the BMI, MetS components, dietary habits, *FTO* polymorphisms, and diplotypes in a population of Polish young adult males.

Supplementary Materials: The following supporting information can be downloaded at: https://www.mdpi.com/article/10.3390/nu16111615/s1, Table S1: polymorphisms information; Table S2: Genotypes frequencies of *FTO* polymorphisms; Table S3: Haplotype frequencies; Table S4: Diplotype frequencies; Table S5: PCA-driven dietary patterns (DPs) identified in the study population (n = 278) by principal component analysis: data from FFQ-6 questionnaire; Table S6: Level HBA1c, BMI category, and familiar history of DM in relation to *FTO* diplotypes (three most common haplotype pairs).

Author Contributions: Conceptualization, T.O. and S.G.-K.; methodology T.O. and S.G.-K.; validation, M.F. and B.M.; formal analysis, M.L.; investigation, M.L.; resources, M.B.; data curation, M.L.; writing—original draft preparation, M.L. and M.G.; writing—review and editing, S.G.-K., T.O., M.G.

and A.T.; visualization, M.L.; supervision, T.O. and S.G.-K.; project administration, S.G.-K., T.O. and M.L.; funding acquisition, T.O. All authors have read and agreed to the published version of the manuscript.

Funding: The study was financed by the National Science Centre (UMO-2014/13/B/NZ5/03166, OPUS 7).

Institutional Review Board Statement: The study protocol has been approved by the ethics committee of the Institute of Occupational Medicine and Environmental Health, Sosnowiec (Resolution no. 03/2013) and the Bioethics Committee of Medical University of Silesia (Permission No. KNW/022/KB1/118/18/19) and conformed to the Declaration of Helsinki.

Informed Consent Statement: Informed consent was obtained from all subjects involved in the study.

Data Availability Statement: The data presented in this study are available on request from the corresponding author. The data are not publicly available due to the planned preparation of subsequent publications based on the collected dataset (data may be publicly available after the end of the project, currently only upon reasonable request).

Conflicts of Interest: The authors declare no conflicts of interest.

References

1. Obesity. Available online: https://www.who.int/health-topics/obesity (accessed on 29 September 2023).
2. Guh, D.P.; Zhang, W.; Bansback, N.; Amarsi, Z.; Birmingham, C.L.; Anis, A.H. The Incidence of Co-Morbidities Related to Obesity and Overweight: A Systematic Review and Meta-Analysis. *BMC Public Health* **2009**, *9*, 88. [CrossRef] [PubMed]
3. Dobrowolski, P.; Prejbisz, A.; Kuryłowicz, A.; Baska, A.; Burchardt, P.; Chlebus, K.; Dzida, G.; Jankowski, P.; Jaroszewicz, J.; Jaworski, P.; et al. Metabolic Syndrome—A New Definition and Management guidelinesA Joint Position Paper by the Polish Society of Hypertension, Polish Society for the Treatment of Obesity, Polish Lipid Association, Polish Association for Study of Liver, Polish Society of Family Medicine, Polish Society of Lifestyle Medicine, Division of Prevention and Epidemiology Polish Cardiac Society, "Club 30" Polish Cardiac Society, and Division of Metabolic and Bariatric Surgery Society of Polish Surgeons. *Arch. Med. Sci.* **2022**, *18*, 1133–1156. [CrossRef]
4. Do Vale Moreira, N.C.; Hussain, A.; Bhowmik, B.; Mdala, I.; Siddiquee, T.; Fernandes, V.O.; Montenegro Júnior, R.M.; Meyer, H.E. Prevalence of Metabolic Syndrome by Different Definitions, and Its Association with Type 2 Diabetes, Pre-Diabetes, and Cardiovascular Disease Risk in Brazil. *Diabetes Metab. Syndr. Clin. Res. Rev.* **2020**, *14*, 1217–1224. [CrossRef] [PubMed]
5. Alberti, K.G.M.M.; Eckel, R.H.; Grundy, S.M.; Zimmet, P.Z.; Cleeman, J.I.; Donato, K.A.; Fruchart, J.-C.; James, W.P.T.; Loria, C.M.; Smith, S.C. Harmonizing the Metabolic Syndrome: A Joint Interim Statement of the International Diabetes Federation Task Force on Epidemiology and Prevention; National Heart, Lung, and Blood Institute; American Heart Association; World Heart Federation; International Atherosclerosis Society; and International Association for the Study of Obesity. *Circulation* **2009**, *120*, 1640–1645. [CrossRef] [PubMed]
6. Osadnik, K.; Osadnik, T.; Gierlotka, M.; Windak, A.; Tomasik, T.; Mastej, M.; Kuras, A.; Jóźwiak, K.; Penson, P.E.; Lip, G.Y.H.; et al. Metabolic Syndrome Is Associated with Similar Long-Term Prognosis in Those Living with and without Obesity: An Analysis of 45 615 Patients from the Nationwide LIPIDOGRAM 2004–2015 Studies. *Eur. J. Prev. Cardiol.* **2023**, *30*, 1195–1204. [CrossRef] [PubMed]
7. Racette, S.B.; Deusinger, S.S.; Deusinger, R.H. Obesity: Overview of Prevalence, Etiology, and Treatment. *Phys. Ther.* **2003**, *83*, 276–288. [CrossRef] [PubMed]
8. Elks, C.E.; Den Hoed, M.; Zhao, J.H.; Sharp, S.J.; Wareham, N.J.; Loos, R.J.F.; Ong, K.K. Variability in the Heritability of Body Mass Index: A Systematic Review and Meta-Regression. *Front. Endocrin.* **2012**, *3*, 29. [CrossRef] [PubMed]
9. Maes, H.H.M.; Neale, M.C.; Eaves, L.J. Genetic and Environmental Factors in Relative Body Weight and Human Adiposity. *Behav. Genet.* **1997**, *27*, 325–351. [CrossRef] [PubMed]
10. Loos, R.J.F.; Yeo, G.S.H. The Genetics of Obesity: From Discovery to Biology. *Nat. Rev. Genet.* **2022**, *23*, 120–133. [CrossRef]
11. Huvenne, H.; Dubern, B.; Clément, K.; Poitou, C. Rare Genetic Forms of Obesity: Clinical Approach and Current Treatments in 2016. *Obes Facts* **2016**, *9*, 158–173. [CrossRef]
12. Frayling, T.M.; Timpson, N.J.; Weedon, M.N.; Zeggini, E.; Freathy, R.M.; Lindgren, C.M.; Perry, J.R.B.; Elliott, K.S.; Lango, H.; Rayner, N.W.; et al. A Common Variant in the *FTO* Gene Is Associated with Body Mass Index and Predisposes to Childhood and Adult Obesity. *Science* **2007**, *316*, 889–894. [CrossRef] [PubMed]
13. Larder, R.; Cheung, M.K.M.; Tung, Y.C.L.; Yeo, G.S.H.; Coll, A.P. Where to Go with *FTO*? *Trends Endocrinol. Metab.* **2011**, *22*, 53–59. [CrossRef] [PubMed]
14. Jia, G.; Fu, Y.; Zhao, X.; Dai, Q.; Zheng, G.; Yang, Y.; Yi, C.; Lindahl, T.; Pan, T.; Yang, Y.-G.; et al. N6-Methyladenosine in Nuclear RNA Is a Major Substrate of the Obesity-Associated *FTO*. *Nat. Chem. Biol.* **2011**, *7*, 885–887. [CrossRef] [PubMed]
15. Berulava, T.; Horsthemke, B. The Obesity-Associated SNPs in Intron 1 of the *FTO* Gene Affect Primary Transcript Levels. *Eur. J. Hum. Genet.* **2010**, *18*, 1054–1056. [CrossRef] [PubMed]

16. Karra, E.; O'Daly, O.G.; Choudhury, A.I.; Yousseif, A.; Millership, S.; Neary, M.T.; Scott, W.R.; Chandarana, K.; Manning, S.; Hess, M.E.; et al. A Link between *FTO*, Ghrelin, and Impaired Brain Food-Cue Responsivity. *J. Clin. Investig.* **2013**, *123*, 3539–3551. [CrossRef] [PubMed]
17. Zhang, M.; Zhang, Y.; Ma, J.; Guo, F.; Cao, Q.; Zhang, Y.; Zhou, B.; Chai, J.; Zhao, W.; Zhao, R. The Demethylase Activity of *FTO* (Fat Mass and Obesity Associated Protein) Is Required for Preadipocyte Differentiation. *PLoS ONE* **2015**, *10*, e0133788. [CrossRef] [PubMed]
18. Ben-Haim, M.S.; Moshitch-Moshkovitz, S.; Rechavi, G. *FTO*: Linking m6A Demethylation to Adipogenesis. *Cell Res.* **2015**, *25*, 3–4. [CrossRef]
19. Zhao, X.; Yang, Y.; Sun, B.-F.; Shi, Y.; Yang, X.; Xiao, W.; Hao, Y.-J.; Ping, X.-L.; Chen, Y.-S.; Wang, W.-J.; et al. *FTO*-Dependent Demethylation of N6-Methyladenosine Regulates mRNA Splicing and Is Required for Adipogenesis. *Cell Res.* **2014**, *24*, 1403–1419. [CrossRef]
20. Merkestein, M.; Laber, S.; McMurray, F.; Andrew, D.; Sachse, G.; Sanderson, J.; Li, M.; Usher, S.; Sellayah, D.; Ashcroft, F.M.; et al. *FTO* Influences Adipogenesis by Regulating Mitotic Clonal Expansion. *Nat. Commun.* **2015**, *6*, 6792. [CrossRef]
21. Azzam, S.K.; Alsafar, H.; Sajini, A.A. *FTO* m6A Demethylase in Obesity and Cancer: Implications and Underlying Molecular Mechanisms. *IJMS* **2022**, *23*, 3800. [CrossRef]
22. Stratigopoulos, G.; Martin Carli, J.F.; O'Day, D.R.; Wang, L.; LeDuc, C.A.; Lanzano, P.; Chung, W.K.; Rosenbaum, M.; Egli, D.; Doherty, D.A.; et al. Hypomorphism for RPGRIP1L, a Ciliary Gene Vicinal to the *FTO* Locus, Causes Increased Adiposity in Mice. *Cell Metab.* **2014**, *19*, 767–779. [CrossRef] [PubMed]
23. Claussnitzer, M.; Dankel, S.N.; Kim, K.-H.; Quon, G.; Meuleman, W.; Haugen, C.; Glunk, V.; Sousa, I.S.; Beaudry, J.L.; Puviindran, V.; et al. *FTO* Obesity Variant Circuitry and Adipocyte Browning in Humans. *N. Engl. J. Med.* **2015**, *373*, 895–907. [CrossRef] [PubMed]
24. Smemo, S.; Tena, J.J.; Kim, K.-H.; Gamazon, E.R.; Sakabe, N.J.; Gómez-Marín, C.; Aneas, I.; Credidio, F.L.; Sobreira, D.R.; Wasserman, N.F.; et al. Obesity-Associated Variants within *FTO* Form Long-Range Functional Connections with IRX3. *Nature* **2014**, *507*, 371–375. [CrossRef] [PubMed]
25. Doaei, S.; Kalantari, N.; Mohammadi, N.K.; Tabesh, G.A.; Gholamalizadeh, M. Macronutrients and the *FTO* Gene Expression in Hypothalamus; a Systematic Review of Experimental Studies. *Indian Heart J.* **2017**, *69*, 277–281. [CrossRef] [PubMed]
26. Mehrdad, M.; Doaei, S.; Gholamalizadeh, M.; Eftekhari, M.H. The Association between *FTO* Genotype with Macronutrients and Calorie Intake in Overweight Adults. *Lipids Health Dis.* **2020**, *19*, 197. [CrossRef] [PubMed]
27. Olszewski, P.K.; Fredriksson, R.; Olszewska, A.M.; Stephansson, O.; Alsiö, J.; Radomska, K.J.; Levine, A.S.; Schiöth, H.B. Hypothalamic *FTO* Is Associated with the Regulation of Energy Intake Not Feeding Reward. *BMC Neurosci.* **2009**, *10*, 129. [CrossRef] [PubMed]
28. Dina, C.; Meyre, D.; Gallina, S.; Durand, E.; Körner, A.; Jacobson, P.; Carlsson, L.M.S.; Kiess, W.; Vatin, V.; Lecoeur, C.; et al. Variation in *FTO* Contributes to Childhood Obesity and Severe Adult Obesity. *Nat. Genet.* **2007**, *39*, 724–726. [CrossRef] [PubMed]
29. Kolackov, K.; Łaczmański, Ł.; Lwow, F.; Ramsey, D.; Zdrojowy-Wełna, A.; Tupikowska, M.; Bednarek-Tupikowska, G. The Frequencies of Haplotypes of *FTO* Gene Variants and Their Association with the Distribution of Body Fat in Non-Obese Poles. *Adv. Clin. Exp. Med.* **2016**, *25*, 33–42. [CrossRef] [PubMed]
30. Jarczak, J.; Grochowalski, Ł.; Marciniak, B.; Lach, J.; Słomka, M.; Sobalska-Kwapis, M.; Lorkiewicz, W.; Pułaski, Ł.; Strapagiel, D. Mitochondrial DNA Variability of the Polish Population. *Eur. J. Hum. Genet.* **2019**, *27*, 1304–1314. [CrossRef]
31. Osadnik, T.; Osadnik, K.; Pawlas, N.; Strzelczyk, J.; Kasperczyk, J.; Poloński, L.; Gąsior, M. Metabolic and Genetic Profiling of Young Adults with and without a Family History of Premature Coronary Heart Disease (MAGNETIC). Study Design and Methodology. *Arch. Med. Sci.* **2019**, *15*, 590–597. [CrossRef]
32. Niedzwiedzka, E.; Wadolowska, L.; Kowalkowska, J. Reproducibility of A Non-Quantitative Food Frequency Questionnaire (62-Item FFQ-6) and PCA-Driven Dietary Pattern Identification in 13–21-Year-Old Females. *Nutrients* **2019**, *11*, 2183. [CrossRef] [PubMed]
33. Kowalkowska, J.; Wadolowska, L.; Czarnocinska, J.; Czlapka-Matyasik, M.; Galinski, G.; Jezewska-Zychowicz, M.; Bronkowska, M.; Dlugosz, A.; Loboda, D.; Wyka, J. Reproducibility of a Questionnaire for Dietary Habits, Lifestyle and Nutrition Knowledge Assessment (KomPAN) in Polish Adolescents and Adults. *Nutrients* **2018**, *10*, 1845. [CrossRef] [PubMed]
34. Posit. Available online: https://www.posit.co/ (accessed on 29 September 2023).
35. Wigginton, J.E.; Cutler, D.J.; Abecasis, G.R. A Note on Exact Tests of Hardy-Weinberg Equilibrium. *Am. J. Hum. Genet.* **2005**, *76*, 887–893. [CrossRef] [PubMed]
36. González, J.R.; Armengol, L.; Solé, X.; Guinó, E.; Mercader, J.M.; Estivill, X.; Moreno, V. SNPassoc: An R Package to Perform Whole Genome Association Studies. *Bioinformatics* **2007**, *23*, 654–655. [CrossRef] [PubMed]
37. Osadnik, T.; Pawlas, N.; Lonnie, M.; Osadnik, K.; Lejawa, M.; Wądołowska, L.; Bujak, K.; Fronczek, M.; Reguła, R.; Gawlita, M.; et al. Family History of Premature Coronary Artery Disease (P-CAD)—A Non-Modifiable Risk Factor? Dietary Patterns of Young Healthy Offspring of P-CAD Patients: A Case-Control Study (MAGNETIC Project). *Nutrients* **2018**, *10*, 1488. [CrossRef] [PubMed]
38. Osadnik, K.; Osadnik, T.; Lonnie, M.; Lejawa, M.; Reguła, R.; Fronczek, M.; Gawlita, M.; Wądołowska, L.; Gąsior, M.; Pawlas, N. Metabolically Healthy Obese and Metabolic Syndrome of the Lean: The Importance of Diet Quality. Analysis of MAGNETIC Cohort. *Nutr. J.* **2020**, *19*, 19. [CrossRef] [PubMed]

39. Hardy, R.; Wills, A.K.; Wong, A.; Elks, C.E.; Wareham, N.J.; Loos, R.J.F.; Kuh, D.; Ong, K.K. Life Course Variations in the Associations between *FTO* and MC4R Gene Variants and Body Size. *Hum. Mol. Genet.* **2009**, *19*, 545–552. [CrossRef]
40. Wrzosek, M.; Zakrzewska, A.; Ruczko, L.; Jabłonowska-Lietz, B.; Nowicka, G. Association between Rs9930506 Polymorphism of the Fat Mass & Obesity-Associated (*FTO*) Gene & Onset of Obesity in Polish Adults. *Indian J. Med. Res.* **2016**, *143*, 281. [CrossRef] [PubMed]
41. Piwonska, A.M.; Cicha-Mikolajczyk, A.; Sobczyk-Kopciol, A.; Piwonski, J.; Drygas, W.; Kwasniewska, M.; Pajak, A.; Zdrojewski, T.; Tykarski, A.; Kozakiewicz, K.; et al. Independent Association of *FTO* Rs9939609 Polymorphism with Overweight and Obesity in Polish Adults. Results from the Representative Population-Based WOBASZ Study. *J. Physiol. Pharmacol.* **2022**, *73*, 395–402. [CrossRef] [PubMed]
42. Luczynski, W.; Zalewski, G.; Bossowski, A. The Association of the *FTO* Rs9939609 Polymorphism with Obesity and Metabolic Risk Factors for Cardiovascular Diseases in Polish Children. *J. Physiol. Pharmacol.* **2012**, *63*, 241–248.
43. Sobalska-Kwapis, M.; Suchanecka, A.; Słomka, M.; Siewierska-Górska, A.; Kępka, E.; Strapagiel, D. Genetic Association of *FTO*/IRX Region with Obesity and Overweight in the Polish Population. *PLoS ONE* **2017**, *12*, e0180295. [CrossRef] [PubMed]
44. Zdrojowy-Wełna, A.; Bednarek-Tupikowska, G.; Zatońska, K.; Kolackov, K.; Jokiel-Rokita, A.; Bolanowski, M. The Association between *FTO* Gene Polymorphism Rs9939609 and Obesity Is Sex-Specific in the Population of PURE Study in Poland. *Adv. Clin. Exp. Med.* **2020**, *29*, 25–32. [CrossRef] [PubMed]
45. Zdrojowy-Wełna, A.; Ramsey, D.; Kolačkov, K.; Słoka, N.; Zatońska, K.; Szuba, A.; Bolanowski, M. The *FTO* Gene Is Not Associated with Weight Gain during Six Years of Observation in the Population of the PURE Study in Poland. *Endokrynol. Pol.* **2020**, *71*, 376–381. [CrossRef] [PubMed]
46. Guclu-Geyik, F.; Onat, A.; Yuzbasıogulları, A.B.; Coban, N.; Can, G.; Lehtimäki, T.; Erginel-Unaltuna, N. Risk of Obesity and Metabolic Syndrome Associated with *FTO* Gene Variants Discloses Clinically Relevant Gender Difference among Turks. *Mol. Biol. Rep.* **2016**, *43*, 485–494. [CrossRef] [PubMed]
47. Karns, R.; Viali, S.; Tuitele, J.; Sun, G.; Cheng, H.; Weeks, D.E.; McGarvey, S.T.; Deka, R. Common Variants in *FTO* Are Not Significantly Associated with Obesity-Related Phenotypes among Samoans of Polynesia: *FTO* and Obesity among Samoans. *Ann. Hum. Genet.* **2012**, *76*, 17–24. [CrossRef] [PubMed]
48. Leite, L.C.G.; Dos Santos, M.C.; Duarte, N.E.; Horimoto, A.R.V.R.; Crispim, F.; Vieira Filho, J.P.B., Dal Fabbro, A.L.; Franco, L.J.; Moises, R.S. Association of Fat Mass and Obesity-Associated (*FTO*) Gene Rs9939609 with Obesity-Related Traits and Glucose Intolerance in an Indigenous Population, the Xavante. *Diabetes Metab. Syndr. Clin. Res. Rev.* **2022**, *16*, 102358. [CrossRef] [PubMed]
49. Reuter, C.P.; Rosane De Moura Valim, A.; Gaya, A.R.; Borges, T.S.; Klinger, E.I.; Possuelo, L.G.; Franke, S.I.R.; Kmetzsch, L.; Vainstein, M.H.; Prá, D.; et al. *FTO* Polymorphism, Cardiorespiratory Fitness, and Obesity in B Razilian Youth. *Am. J. Hum. Biol.* **2016**, *28*, 381–386. [CrossRef]
50. Jacobsson, J.A.; Risérus, U.; Axelsson, T.; Lannfelt, L.; Schiöth, H.B.; Fredriksson, R. The Common *FTO*variant Rs9939609 Is Not Associated with BMI in a Longitudinal Study on a Cohort of Swedish Men Born 1920–1924. *BMC Med. Genet.* **2009**, *10*, 131. [CrossRef] [PubMed]
51. Harbron, J.; Van der Merwe, L.; Zaahl, M.G.; Kotze, M.J.; Senekal, M. Fat Mass and Obesity-Associated (FTO) Gene Polymorphisms Are Associated with Physical Activity, Food Intake, Eating Behaviors, Psychological Health, and Modeled Change in Body Mass Index in Overweight/Obese Caucasian Adults. *Nutrients* **2014**, *6*, 3130–3152. [CrossRef]
52. Goh, Y.; Choi, J.-H. Genetic Variation Rs1121980 in the Fat Mass and Obesity-Associated Gene (*FTO*) Is Associated with Dietary Intake in Koreans. *Food Nutr. Res.* **2022**, *66*, 8059. [CrossRef]
53. Jacobsson, J.A.; Almén, M.S.; Benedict, C.; Hedberg, L.A.; Michaëlsson, K.; Brooks, S.; Kullberg, J.; Axelsson, T.; Johansson, L.; Ahlström, H.; et al. Detailed Analysis of Variants in *FTO* in Association with Body Composition in a Cohort of 70-Year-Olds Suggests a Weakened Effect among Elderly. *PLoS ONE* **2011**, *6*, e20158. [CrossRef] [PubMed]
54. Wu, J.; Xu, J.; Zhang, Z.; Ren, J.; Li, Y.; Wang, J.; Cao, Y.; Rong, F.; Zhao, R.; Huang, X.; et al. Association of *FTO* Polymorphisms with Obesity and Metabolic Parameters in Han Chinese Adolescents. *PLoS ONE* **2014**, *9*, e98984. [CrossRef] [PubMed]
55. Kalantari, N.; Keshavarz Mohammadi, N.; Izadi, P.; Doaei, S.; Gholamalizadeh, M.; Eini-Zinab, H.; Salonurmi, T.; Rafieifar, S.; Janipoor, R.; Azizi Tabesh, G. A Haplotype of Three SNPs in *FTO* Had a Strong Association with Body Composition and BMI in Iranian Male Adolescents. *PLoS ONE* **2018**, *13*, e0195589. [CrossRef] [PubMed]
56. Freathy, R.M.; Timpson, N.J.; Lawlor, D.A.; Pouta, A.; Ben-Shlomo, Y.; Ruokonen, A.; Ebrahim, S.; Shields, B.; Zeggini, E.; Weedon, M.N.; et al. Common Variation in the *FTO* Gene Alters Diabetes-Related Metabolic Traits to the Extent Expected Given Its Effect on BMI. *Diabetes* **2008**, *57*, 1419–1426. [CrossRef]
57. Doney, A.S.F.; Dannfald, J.; Kimber, C.H.; Donnelly, L.A.; Pearson, E.; Morris, A.D.; Palmer, C.N.A. The *FTO* Gene Is Associated With an Atherogenic Lipid Profile and Myocardial Infarction in Patients With Type 2 Diabetes: A Genetics of Diabetes Audit and Research Study in Tayside Scotland (Go-DARTS) Study. *Circ. Cardiovasc. Genet.* **2009**, *2*, 255–259. [CrossRef]

nutrients

MDPI

Article

Blood Phytosterol Concentration and Genetic Variant Associations in a Sample Population

Leticia Garrido-Sanchez [1,*], Elisabet Leiva-Badosa [1], Josep Llop-Talaveron [1], Xavier Pintó-Sala [2], Toni Lozano-Andreu [3], Emili Corbella-Inglés [2], Pedro Alia-Ramos [4], Lluis Arias-Barquet [5], Josep Maria Ramon-Torrel [6] and Maria B. Badía-Tahull [1]

[1] Pharmacy Department, Hospital Universitari Bellvitge, IDIBELL, Universitat Barcelona, 08907 L'Hospitalet de Llobregat, Spain
[2] Cardiovascular Risk Unit, Internal Medicine, Hospital Universitari de Bellvitge, IDIBELL, Universitat Barcelona, 08907 L'Hospitalet de Llobregat, Spain; xpinto@bellvitgehospital.cat (X.P.-S.)
[3] Pharmacy Department, Institut Català d'Oncologia, IDIBELL, Universitat Barcelona, 08907 L'Hospitalet de Llobregat, Spain
[4] Clinical Laboratory Department, Hospital Universitari Bellvitge, IDIBELL, Universitat Barcelona, 08907 L'Hospitalet de Llobregat, Spain
[5] Ophthalmology Department, Hospital Universitari Bellvitge, IDIBELL, Universitat Barcelona, 08907 L'Hospitalet de Llobregat, Spain
[6] Preventive Medicine Department, Hospital Universitari Bellvitge, IDIBELL, Universitat Barcelona, 08907 L'Hospitalet de Llobregat, Spain
* Correspondence: lgarridos@bellvitgehospital.cat

Abstract: The main objective of this study was to determine plasma levels of PS and to study SNVs rs41360247, rs4245791, rs4148217, and rs11887534 of *ABCG8* and the r657152 SNV at the ABO blood group locus in a sample of a population treated at our hospital, and to determine whether these SNVs are related to plasma PS concentrations. The secondary objective was to establish the variables associated with plasma PS concentrations in adults. Participants completed a dietary habit questionnaire and a blood sample was collected to obtain the following variables: campesterol, sitosterol, sitostanol, lanosterol, stigmasterol, biochemical parameters, and the SNVs. In addition, biometric and demographic variables were also recorded. In the generalized linear model, cholesterol and age were positively associated with total PS levels, while BMI was negatively related. For rs4245791, homozygous T allele individuals showed a significantly lower campesterol concentration compared with C homozygotes, and the GG alleles of rs657152 had the lowest levels of campesterol compared with the other alleles of the SNV. Conclusions: The screening of certain SNVs could help prevent the increase in plasma PS and maybe PNALD in some patients. However, further studies on the determinants of plasma phytosterol concentrations are needed.

Keywords: phytosterols; SNVs; campesterol; sitosterol; ABCG5/8

1. Introduction

Sitosterolaemia or phytosterolaemia is an autosomal recessive disorder of lipid metabolism characterized by phytosterol (PS) accumulation that can lead to atherogenic and cardiovascular disease. Since the first time that phytosterolaemia was described [1], many studies have identified its origin in a mutation in the genes encoding a sterol efflux transporter, *ABCG5/8*; its function is the hepatic secretion and intestinal absorption of cholesterol and plant sterols. Mutations in the genes located on chromosome 2p21 and their genetic variation alter the function of the transporter and, therefore, dietary sterol blood concentrations, mainly vegetal sterols [2–5]. These two proteins, ABCG5 and ABCG8, form a heterodimer of the ATP-binding cassette (ABC) family and are in the canalicular membrane of hepatocytes and the apical membrane of enterocytes. ABCG5/8 is also responsible for the classic reverse cholesterol transport and transintestinal cholesterol excretion [6].

In 2010, a genome-wide association study [7] identified significant associations of plasma plant sterol concentrations with three single nucleotide variants (SNVs): rs4245791, rs41360247, and rs657152. rs4245791 and rs41360247 are genetic variants of the *ABCG8* hemitransporter gene. rs657152, at the blood group ABO locus, was linked to the group O allele and was associated with decreased plasma PS levels. In addition, alleles associated with increased PS levels displayed a significant association with an increased probability of coronary artery disease (CAD), and alleles associated with reduced PS levels were associated with reduced CAD risk [7].

Two other genetic variants, rs4148217 and rs11887534 were associated with blood PS concentrations in previous studies [8–10] (Table 1).

Table 1. Genetic control: variants, alleles, and effects.

Variants (Gene)	Ref Allele > Alt allele (Allele Frequency) [11] *	Effect of Alt Allele on Plasma PS Levels	References
rs41360247 (*ABCG8*)	T > C (C = 0.0604)	Lower levels	[7]
rs4245791 (*ABCG8*)	C > T (C = 0.3195)	Higher levels	[7]
rs4148217 (*ABCG8*)	C > A, T (A = 0.1901)	Lower levels	[8,9,12]
rs11887534 (*ABCG8*)	G > C (C = 0.0651)	Lower levels	[8,10,12]
rs657152 (*ABO*)	A > C, T (A = 0.3915)	Higher levels	[7]

* Population: European (non-Finnish); Ref: reference; Alt: alternative; PS: phytosterols.

Phytosterolaemia is considered an extremely rare disorder; nevertheless, high PS levels may occur with the use of intravenous vegetable-based lipid emulsions (LEs) in patients on parenteral nutrition (PN). Several studies have shown, especially in newborns, that patients treated with PN have higher PS levels than healthy controls [13–16].

Parenteral nutrition-associated liver disease (PNALD), manifesting as liver function test (LFT) alterations, is a relevant complication associated with elevated plasma PS. Our group carried out a randomized double-blind clinical study with 19 patients on PN with vegetable-based LEs. We observed that plasma accumulation of PS and altered LFTs could be prevented with the exclusive administration of plant sterol-free LEs, such as Fish Oil-based LEs [17].

A substudy of the previous clinical trial (EudraCT Number: 2014-003597-17) showed that, depending on the allele pairing of both rs41360247 and rs4245791 SNVs, the LFT alteration in adult patients on short-term PN is conditional. We also concluded that further studies in larger series are necessary to determine the conditions under which the variants act and how they do so [18].

Therefore, genotyping the variants associated with plasma PS levels may allow us to personalize therapeutic strategies for patients treated with PN. In this context, we performed an observational study of a sample population attending our hospital to determine the prevalence of these SNVs and their relationship with plasma PS concentrations.

The main objective of this study was to measure plasma PS concentrations and to study SNVs rs41360247, rs4245791, rs4148217, and rs11887534 of *ABCG8* and the r657152 SNV at the ABO blood group locus in a sample population to determine whether these SNVs are related to plasma PS concentrations. The secondary objective was to establish the variables associated with plasma PS concentrations in adults.

2. Materials and Methods

2.1. The Study Population

The present work is an observational study conducted in an adult population recruited from the Departments of Ophthalmology and Preventive Medicine of Bellvitge University Hospital (L´Hospitalet de Llobregat, Barcelona, Spain). A total of 185 participants [18–81 years old; 94% Caucasian, 6% Latin] completed a dietary habit questionnaire, and blood samples were collected. Patients classified as chronic complex or those with advanced chronic disease were excluded. All volunteers provided written informed consent. The investigation project (PR 144/17) was approved by the Clinical Research Ethics Committee of the hospital on 25 May 2017.

2.2. Data Collection

The collected dependent variables were total PS, campesterol, sitosterol, sitostanol, lanosterol, and stigmasterol. Independent variables were the non-coding variants rs41360247 and rs4245791 of *ABCG8*, the missense variants rs11887534 and rs4148217 of *ABCG8*, and lastly the non-coding variant rs657152 of ABO, as well as the biochemical parameters: total cholesterol, triglycerides, α-tocopherol, gamma-glutamyl transferase (GGT), alkaline phosphatase (AP), alanine transferase (ALT), total bilirubin, creatinine, and urea. In addition, the following biometric and demographic variables were also recorded: sex, age, height (m), weight (Kg), and calculated body mass index (BMI) (kg/m^2).

Dietary habits were collected with a questionnaire (Table A1) that included type of diet and eating habits (type of oil consumed, vegetarian/vegan, and the number of nuts, pastries, sausages, and butter/margarine consumed during the week) and intake of vitamins and PS-enriched supplements [19]. From the data obtained through the dietary questionnaire, the variable fat intake index (FI) was calculated, which included the frequency of saturated fats ingested (butter/margarine, sausages, and pastries) during the week, and scored from 0 (no saturated fats ingested) to 6, the maximum saturated fat intake (>3 times per week of three foods). The vegetal intake index (VII) was also a calculated variable following the same criteria, which included the vegetal ingested (nuts, PS-enriched supplements, and vegetarian diet). Finally, cholesterol-lowering treatments were also registered.

Blood samples were taken after an overnight fast, obtaining plasma values of the following data: total PS and their fractions (campesterol, beta-sitosterol, stigmasterol, sitostanol, and lanosterol); biochemical parameters (total cholesterol, triglycerides, α-tocopherol, LFT (GGT, AP, ALT, and total bilirubin)); renal function parameters (creatinine and urea); and genetic variants (rs11887534, rs4245791, rs41360247, rs4148217, and rs657152).

2.3. Analytical Determinations

- Phytosterol analysis

To determine plasma PS, blood samples were collected in 4 mL tubes of lithium heparin, kept cold at 2–8 °C for up to one hour, and then centrifuged at 2000× *g* for 10 min at 4 °C. Plasma was aliquoted into 5 mL plastic tubes and stored at −80 °C until processing. Measurements of different PS concentrations in the plasma were carried out using the UPLC-ACQUITY TQD measurement system, which uses liquid chromatography of high and rapid resolution (UPLC) coupled to tandem mass spectrometry as a measurement principle (MS/MS). We worked in the reverse-phase modality using a C18 UPLC column that allowed a faster and higher resolution of the chromatographic peaks. The mobile phase was composed of two solutions of ammonium acetate and 0.1% (*v*/*v*) formic acid, one in acetonitrile and the other in methanol, using a gradient elution. As a quality control, all the samples were two-fold analyzed [17].

- α-tocopherol analysis

To determine plasma α-tocopherol, blood samples were collected in 4 mL tubes of lithium heparin and kept cold at 2–8 °C for up to one hour. They were centrifuged at 2000× *g* for 10 min at 4 °C. Measurements of α-tocopherol concentrations in the plasma were carried out using the UPLC-ACQUITY TQD. The analytical variation (CVs) was between 6.1% (for a mean of 26.2 mmol/L) and 3.7% (for a mean of 63.7 mmol/L).

- DNA isolation and genotyping

Blood samples from all individuals were collected in tubes containing ethylenediaminetetraacetic acid, and genomic DNA was isolated from peripheral blood leukocytes by using the automated format Maxwell 16 (Promega, Madison, WI, USA) Blood DNA Purification Kit. All DNA samples were stored at 4 °C until polymerase chain reaction (PCR) applications were performed.

TaqMan assays (ThermoFisher) were used according to the manufacturer's instructions to determine genotypes for variants rs11887534 (C__26135643_10), rs4148217 (C____375061_10), and rs41360247 (C__86448255_10). To determine genotypes for variants rs4245791 and rs657152, PCR-restriction fragment length polymorphism (RFLP) analyses were performed. Briefly, in the case of rs4245791, DNA was PCR amplified using primers: forward 5′-CGTCTGGTAGATAAGTTCTGGT-3′, in which the last G substitutes a T to create a restriction site for BstEII; and reverse: 5′-CTGGCCGGGATCTACTTTT-3′. For variant rs657152, primers were forward 5′-GCAGAATGGCTGAGAACACA-3′ and reverse 5′-TACATGCTGGAGCTGTTTGC-3′, and the amplified 195-fragment was cut with MseI (both restriction enzymes were purchased from New England Biolabs, Ipswich, MA, USA). DNA integrity is controlled by agarose gel electrophoresis; DNA concentration mean range: 40–100 ng/mL.

2.4. Statistical Analysis

We calculated that a random sample of 203 subjects could be enough to estimate, with a confidence of 95% and a precision of ±5%, considering the frequency of the minor allele of rs4245791 (this variant was the main objective in our initial study) [7]. The percentage of necessary replacements was expected to be 20%.

A descriptive statistical analysis was performed using frequency tables for all variables. For continuous variables, descriptive parameters such as *n*, mean, and standard deviation (SD) were used. For categorical variables, grouped percentages were given and a chi-square analysis was carried out. The sample was stratified based on the percentile 50 and a comparison was made between the two groups above and below the percentile 50. To evaluate the presence of an association between the categorical variables (nuts, pastries, and sausages) and PS, we incorporated the linear-by-linear association (LLA) test in the Crosstab. Simple linear regression tests were performed for continuous variables. To study the relationship between plasma concentrations of PS and fractions and variants, a one-way ANOVA analysis was performed.

Finally, three generalized linear models were run, one for each of the following dependent variables: total PS, sitosterol, and campesterol, including the independent variables and using a linear equation minimizing residual deviance. In the case of heteroscedasticity in the residual of the model, the robust estimator was used. The overall significance of the model was established with the Omnibus test χ^2. Models without statistical significance were refused. As complementary information, in all three models, a Bonferroni simultaneous multiple comparison test was performed.

Data were analyzed using IBM SPSS 28.0; statistical significance was reported with a 95% confidence interval (CI) at the conventional $p < 0.05$ (two-tailed).

3. Results

A group of 185 volunteers were recruited for the study. Plasma levels of PS were (mean ± SD): total phytosterols 1.94 ± 1.48 µcg/mL, campesterol 0.85 ± 0.91 µcg/mL, sitosterol 0.85 ± 0.73 µcg/mL, stigmasterol 0.10 ± 0.13 µcg/mL, lanosterol 0.09 ± 0.09 µcg/mL, and sitostanol 0.02 ± 0.02 µcg/mL.

Table 2 shows the demographic, biometric, dietary, and analytical data of the study individuals separated into two groups: one with PS levels above and the other below the median (P50).

The variant frequency of the population studied is shown in Table 3. All genotype frequency distributions were in Hardy–Weinberg equilibrium.

Of the five SNVs studied, only rs4245791, rs4148217, and rs657152 had data for the minor homozygous alleles. Total cholesterol, total PS, and campesterol plasma concentrations are shown in Figure 1. Differences between PS and fractions and SNVs rs41360247 and rs11887543 were not statistically significant (Table S1 Supplementary Materials).

Table 2. Demographic, biometric, dietary, and analytical data of the studied subjects separated by median PS.

Variable	Total (n = 185) Mean ± SD/N (%)	PS ≤ P50 (n = 93) Mean ± SD/N (%)	PS > P50 (n = 92) Mean ± SD/N (%)	p *
Demographic and anthropometric data				
Age (years)	41.49 ± 18.57	42.82 ± 19.07	40.14 ± 18.07	0.164
BMI (kg/m^2)	24.47 ± 4.28	25.23 ± 4.83	23.71 ± 3.51	**0.008**
Sex (women)	115 (62.20%)	56 (60.20%)	59 (64.10%)	0.583
Dietary habits				
Vegetarian-vegan (yes)	9 (4.90%)	4 (4.30%)	5 (5.40%)	0.720
Nuts				0.125 **
No consumption	58 (31.40%)	32 (34.4%)	26 (28.30%)	
1–3 times a week	99 (53.50%)	51 (54.80%)	48 (52.20%)	
>3 times a week	28 (15.10%)	10 (10.80%)	18 (19.60%)	
Pastries				0.069 **
No consumption	94 (50.80%)	43 (46.20%)	51 (55.40%)	
1–3 times a week	78 (42.20%)	40 (43.00%)	38 (41.30%)	
>3 times a week	13 (7.00%)	10 (10.80%)	3 (3.30%)	
Sausages				0.632 **
No consumption	24 (13.00%)	13 (14.00%)	11 (12.00%)	
1–3 times a week	103 (55.70%)	52 (55.90%)	51 (55.40%)	
>3 times a week	58 (31.40%)	28 (30.10%)	30 (32.60%)	
FI (0–6)	2.05 ± 1.16	2.06 ± 1.24	2.04 ± 1.07	0.451
VII (0–6)	1.22 ± 1.33	0.98 ± 1.11	1.34 ± 1.46	**0.031**
Cholesterol treatment and dietary supplementation				
Cholesterol-lowering treatment (Yes)	16 (8.6%)	9 (9.7%)	7 (7.6%)	0.617
PS-enriched supplements (Yes)	8 (4.30%)	2 (2.2%)	6 (6.5%)	0.144
Vitamins (Yes)	12 (6.50%)	6 (6.50%)	6 (6.50%)	0.985
Plasma values				
Total cholesterol (mg/dL)	101.92 ± 57.65	88.18 ± 54.63	115.81 ± 57.57	**<0.001**
Alanine aminotransferase (μKat/L)	0.29 ± 0.13	0.29 ± 0.14	0.28 ± 0.12	0.323
Alkaline phosphatase (μKat/L)	0.98 ± 0.28	1.00 ± 0.30	0.96 ± 0.25	0.184
α-Tocopherol (μmol/L)	30.25 ± 6.20	30.12 ± 5.99	30.39 ± 6.42	0.380
Creatinine (μmol/L)	76.82 ± 16.83	78.24 ± 17.67	75.38 ± 15.89	0.125
Albumin (g/L)	47.74 ± 3.29	47.82 ± 3.34	47.66 ± 3.26	0.375

FI: fat intake index; VII: vegetal intake index; PS > PS50: phytosterol concentrations above the median; PS ≤ P50: phytosterol concentrations less than or equal to the median; * differences between PS ≤ P50 and PS > P50; ** statistical significance for the linear trend measure in the PS > P50 group vs. PS ≤ P50 group.

The relation between PS, categorical, and continuous variables is depicted in Table S2a,b (Supplementary Materials).

In the generalized linear model, cholesterol and age were positively associated with total PS, while BMI was negatively associated. Also, α-Tocopherol tended towards a negative association with campesterol (Table 4).

There were no homozygous minor allele carriers for rs41360247 and rs11887534, so these variants were excluded from the multivariant analysis. For rs4245791, T homozygotes had significantly lower campesterol concentrations compared with C homozygotes. For rs4148217, CA heterozygotes presented lower sitosterol and total PS levels versus major allele homozygotes (CC), as a trend. Finally, GG homozygotes of rs657152 had the lowest levels of campesterol compared with the other alleles of the SNV.

Table 3. Single nucleotide variant frequency of the *ABCG8* gene and the blood group ABO locus in the study population.

Polymorphism	Genotypes (N/%)		
rs41360247	CC 0 (0%)	TC 26 (14.1%)	TT 159 (85.9%)
rs4245791	CC 19 (10.3%)	TC 79 (42.7%)	TT 87 (47.0%)
rs4148217	AA 5 (2.7%)	CA 58 (31.4%)	CC 122 (65.9%)
rs11887534	CC 0 (0%)	GC 27 (14.6%)	GG 158 (85.4%)
rs657152	TT 29 (15.7%)	GT 90 (48.6%)	GG 66 (35.7%)

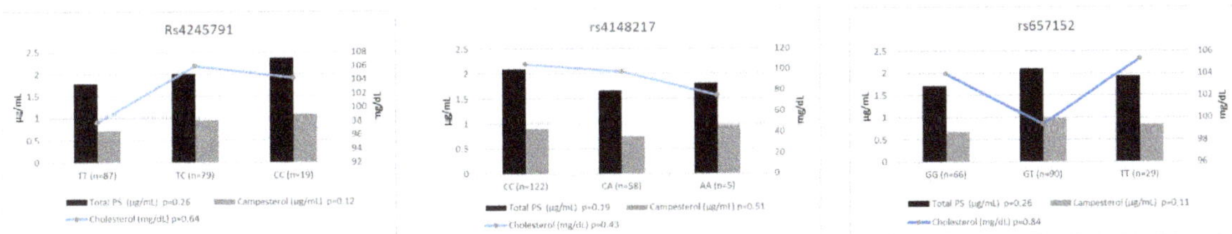

Figure 1. Total cholesterol, total phytosterols, and campesterol plasma concentrations for rs4245791, rs4148217, and rs657152.

Table 4. General linear model: PS and variables.

Variable	Total PS		Sitosterol		Campesterol	
	B [95% CI]	*p*	B [95% CI]	*p*	B [95% CI]	*p*
Cholesterol (mg/dL) (×10⁴)	10.00 [0.00–10.00]	**<0.001**	0.00 [0.00–10.00]	**<0.001**	0.00 [−0.18–10]	0.065
rs4245791_TT vs. rs4245791_CC rs4245791_TC vs. rs4245791_CC	−0.57 [−1.28–0.15] −0.41 [−1.10–0.28]	0.119 0.244	−0.16 [−0.47–0.15] −0.23 [−0.52–0.07]	0.311 0.137	−0.48 [−0.93–(−0.03)] −0.20 [−0.63–0.22]	**0.037** 0.344
rs4148217_AA vs. rs4148217_CC rs4148217_AC vs. rs4148217_CC	0.10 [−1.16–1.36] −0.38 [−0.83–0.06]	0.880 0.093	−0.16 [−0.70–0.39] −0.19 [−0.38–0.01]	0.577 0.059	0.40 [−0.33–1.13] −0.12 [−0.42–0.18]	0.282 0.443
rs6577152_TT vs. rs6577152_GG rs6577152_GT vs. rs6577152_GG	0.29 [−0.34–0.92] 0.52 [0.07–0.96]	0.366 **0.022**	0.08 [−0.20–0.35] 0.17 [−0.02–0.36]	0.584 0.074	0.26 [0.02–0.51] 0.36 [0.07–0.65]	**0.035** **0.016**
Sex (women)	0.20 [−0.34–0.74]	0.477	0.19 [−0.04–0.42]	0.109	−0.02 [−0.43–0.38]	0.907
Age (years) (×10)	0.23 [0.09–0.37]	**0.002**	0.02 [−0.04–0.08]	0.491	0.17 [0.09–0.25]	**<0.001**
BMI (kg/m²) (×10)	−0.57 [−1.10–(−0.04)]	**0.034**	−0.25 [−0.47–(−0.02)]	**0.033**	−0.37 [−0.69–(−0.04)]	**0.029**
Cholesterol-lowering treatment (Yes)	−0.10 [−0.94–0.74]	0.813	0.05 [−0.31–0.41]	0.791	0.04 [−0.43–0.51]	0.870
Vitamins (Yes)	−0.17 [−1.00–0.65]	0.685	0.04 [−0.32–0.40]	0.827	−0.19 [−0.58–0.20]	0.349
Alanine aminotransferase (μKat/L)	0.22 [−1.45–1.90]	0.794	0.27 [−0.46–1.00]	0.468	−0.20 [−0.89–0.49]	0.566
Alkaline phosphatase (μKat/L)	0.58 [−0.22–1.38]	0.152	0.12 [−0.23–0.46]	0.512	0.50 [−0.34–1.33]	0.247
Creatinine (μmol/L) (×10⁴)	−80.00 [−220.00–70.00]	0.303	0.01 [−60.00–60.00]	1.000	−80.00 [−180.00–20.00]	0.127
FI	−0.06 [−0.24–0.12]	0.494	−0.04 [−0.12–0.04]	0.300	−0.04 [−0.16–0.09]	0.551
VII	0.03 [−0.13–0.18]	0.755	0.03 [−0.03–0.10]	0.334	−0.02 [−0.10–0.07]	0.723
α-Tocopherol (μmol/L) (×10)	−0.21 [−0.55–0.13]	0.222	−0.02 [−0.17–0.12]	0.757	−0.19 [−0.39–0.01]	0.057
Omnibus χ^2	41.38	**<0.001**	54.50	**<0.001**	43.95	**<0.001**
R^2	0.20		0.21		0.25	

FI: fat intake index; VII: vegetal intake index.

In the Bonferroni simultaneous multiple test, the comparison of blood PS concentrations and genetic variants displays a similar trend toward statistical significance (Table S3 Supplementary Materials).

4. Discussion

Genetic approach: SNVs and PS: Our study evidences a relationship between some variants in the genes encoding ABCG8 and the blood group ABO locus and PS plasma levels. This supports the hypothesis that PS levels are under tight genetic control. In previous studies from other authors, a genome-wide association showed that campesterol and sitosterol were higher in rs41360247 major allele homozygote and rs4245791 [20] and rs657152 [7] minor allele homozygotes. Silbernagel [21] described similar data, and three more SNVs (rs4299376, rs6576629, and rs4953023), not included in our study, were also studied.

In other studies, the minor allele carriers for rs11887534 showed lower plasma campesterol, sitosterol [22], LDL, and total cholesterol levels than wild-type subjects [10]. The sitosterol to cholesterol ratio for rs4148217 (T400K) minor allele carriers was lower than wild type [8]. Plat et al. found that the association between variants and PS levels was allele-dependent. Moreover, subjects with higher PS concentrations presented the homozygous wild-type rs4148217 genotype and had the greatest reductions in sitosterol levels after four weeks of plant stanol ester consumption [9] compared with carriers of minor alleles. Helgadottir´s study found an association between rs4148217, rs11887534, PS levels, and the risk of CAD. Nine variants of *ABCG5/8* were associated with PS levels [12].

In our series, as in the studies described, homozygous T carriers of rs4245791 had lower levels of PS, reaching significant differences for campesterol; for rs657152, campesterol levels in GG individuals were significantly lower than those in subjects with mutant alleles.

The SNV rs4148217 only showed a tendency toward a lower plasma sitosterol concentration in heterozygous individuals. Homozygous minor allele carriers had higher levels of campesterol than the others, the reverse of previous reports. This effect could be explained by the limited number of subjects and the presence of outlier data in the AA group.

Hepatic function and PS: Case-control, randomized cross-over intervention, or cross-sectional studies were performed to study the relation between PS and coronary heart disease (CHD) risk. Different results were reported: a significant association in some cases, no relation in others, and a protective effect of PS on CHD development was also described [23–28].

While initial studies tried to demonstrate the association between PS and CHD, an association between PS and liver function alterations has also been reported under certain conditions. In fact, high PS levels have been reported in patients receiving PN due to the administration of vegetable oil-based LEs. Ellegard et al. quantified the plasma PS levels of 21 healthy adult controls and 24 adult patients with short bowel syndrome receiving PN or not. Mean plasma PS levels were 23 μmol/L in controls, 11 μmol/L in short-bowel patients not receiving PN, and 63 μmol/L in short-bowel patients receiving PN [29]. Our group published a study of twenty-seven adult intestinal failure patients on home PN vs. seven adult controls and found PS levels of 55.4 ± 6.2 μg/mL vs. 14.8 ± 2.3 μg/mL, respectively; all LFT variables studied showed a statistical association with plasma PS in patients on PN [30]. GGT and ALT increase the association with PS in adults on PN and hepatic alterations were also seen [17]. In our data, ALT and AP did not show significant differences in the multivariate model.

Concerning SNVs, data obtained from a substudy in PN patients found that increases in AP were associated with the T allele of rs41360247 [18]. We must consider that subjects in our sample population were not on PN and had LFTs within the normal reference range.

It is also relevant to highlight that plasma α-tocopherol tends towards an inverse association with campesterol levels; however, Gylling found a positive association between plant sterols and α-tocopherol, suggesting that higher cholesterol absorption efficiency can also raise plasmatic α-tocopherol levels [31]. On the other hand, the antioxidant effect of

α-tocopherol can be useful in treating non-alcoholic steatohepatitis [32], and the addition of α-tocopherol to LEs has been applied to avoid peroxidation due to polyunsaturated fats. We included α-tocopherol in the analysis for its hepatic protection; however, subjects did not present hepatic disease, thus no conclusions can be drawn [33,34].

Demographic and nutritional factors: We found a positive association between age and total PS and campesterol, as shown in some articles. Other factors related to plasma plant sterol levels could be diet, availability of transport vehicles, and hepatic uptake [24,35].

The principal dietary sources of PS are vegetable-based oils, containing higher concentrations by weight than other oils, fruit, vegetables, nuts, and cereals. Many factors affect its bioavailability and bioactivity: sterol type (saturated or glycosylated), source, chemical structure, cooking, and other food ingredients [36]. Related to dietary habits, sitosterol adjusted by cholesterol has shown gradual increases with PS intake [37]. We conducted a dietary questionnaire of the volunteers to include the contribution of diet to PS levels. The ability of PS to reduce LDL levels [38–40] leads to functional foods with plant sterols/stanols and it could be useful in some situations [41]; therefore, for this reason, we have also included stanol-enriched food intake in the dietary questionnaire.

In patients with phytosterolaemia, sitosterol concentrations can reach >80 μg/mL [42,43]. None of the volunteers presented phytosterolaemia, the maximum sitosterol concentrations were 5.73 μg/mL in women and 2.63 μg/mL in men. A higher PS-level association with women had been previously demonstrated [43,44].

With a view to evaluate the effect of PS on the lipid profile, cardiovascular risk, and metabolic syndrome, many studies have determined that plasma campesterol and sitosterol were associated with PS intake, high absorption efficiency of cholesterol, and were inversely correlated with BMI, plasma glucose, and triglycerides [10,37]. In our study, BMI presents a negative association with PS, in addition, the group of volunteers with PS levels below the median had 1.52 kg/m^2 higher BMI with respect PS $\leq$ P50 group.

Cholesterol: The low intestinal absorption rate of PS (<5%), in contrast with cholesterol absorption (50%) [45], also depends on the PS fraction. Campesterol and campestanol have a higher absorption rate than sitosterol (9.6% and 12.5%, respectively) [42]. In fact, we found that plasma total cholesterol was 500 times the value of total PS (2.01 vs. 103.1 mg/dL). Campesterol and sitosterol were the main PS fractions, while sitostanol was the minor fraction and its concentration was stable among variants.

Plasma campesterol and sitosterol levels had a positive correlation with cholesterol absorption, dietary plant sterols, and biliary cholesterol secretion, and are inversely related to cholesterol synthesis [46]. In our study, a positive association between total PS and cholesterol plasma levels, which can be explained through the absorption mechanism shared by both compounds, was found. In addition, the difference between the mean cholesterol in the groups with PS levels above and below the median (P50) was 27.6 mg/dL.

Limitations: This work is our first approach towards a target of adult patients to determine risk factors that can be considered in the clinical field, so the study has some limitations that should be considered for future studies. The limited number of subjects did not allow us to obtain data from homozygous minor allele individuals in two of the five SNVs studied (rs41360247 and rs11887534), and for the other variants, the number of minor allele homozygotes was insufficient to obtain strong associations.

The analytical methods and biological variation are essential, and become important for comparing studies [44].

We collected race as a variable, but 94% of the subjects were Caucasian and the rest Latin; for this reason, we decided to exclude race in the statistical analysis.

With the aim of determining dietary habits, volunteers filled out a survey with no follow-up, thus without intervention on the diet we cannot conclude that the statistically significant difference may be due to the consumption of a food or its absence. It is also relevant to highlight that we did not register the amounts of plant sterols, fat intake, and cholesterol-lowering therapy doses. For this reason, the results obtained can only be considered as trends.

5. Conclusions

Cholesterol, age, and BMI were associated with blood levels of PS. TT individuals for rs4245791 in *ABCG8* and GG individuals for rs657152 in ABO genes display lower blood campesterol levels. For rs4148217 in the ABCG8 gene, AC individuals tended towards lower sitosterol and total PS levels. The screening of certain SNVs could help prevent the increase in plasma PS and maybe PNALD in candidates for long-term PN programs without gastrointestinal stimuli. Further studies on the determinants of plasma phytosterol concentrations are needed.

Supplementary Materials: The following supporting information can be downloaded at https://www.mdpi.com/article/10.3390/nu16071067/s1. (S1. Plasma Sterol levels and variants, S2a. Association between plasma phytosterols and categorical variables (univariant analysis), S2b. Association between plasma phytosterols and continuous variables (univariant analysis), S3. Bonferroni simultaneous multiple comparison test: Total phystosterol, campesterol and sitosterol for each variant (rs424579, rs4148217PS and rs6577152)).

Author Contributions: We certify that all authors have made substantial contributions to the manuscript and thus meet the criteria for authorship. E.L.-B., J.L.-T., and M.B.B.-T. conceived the study and its design; E.L.-B. and T.L.-A. coordinated and performed information collection; J.L.-T. made the statistical analysis; J.L.-T., E.L.-B. and L.G.-S. wrote the manuscript; L.A.-B. and J.M.R.-T. contributed to the provision of patients; and finally, P.A.-R., X.P.-S. and E.C.-I. contributed to practical performance and critical review of the manuscript. All authors have read and agreed to the published version of the manuscript.

Funding: This research received no external funding.

Institutional Review Board Statement: The study was conducted in accordance with the Declaration of Helsinki, and approved by the Clinical Research Ethics Committee of the Hospital Universitari de Bellvitge (PR144/17, 25 May 2017).

Informed Consent Statement: Informed consent was obtained from all subjects involved in the study.

Data Availability Statement: Data is contained within the article (and Supplementary Materials).

Acknowledgments: We thank CERCA programme/Generalitat de Catalunya for institutional support.

Conflicts of Interest: The authors declare no conflicts of interest.

Appendix A

Table A1. Annex I. Dietary habits questionnaire.

1. Are you vegetarian/vegan?	☐ Yes ☐ No
2. What types of oil do you usually consume?	☐ None ☐ Sesame ☐ Olive ☐ Rice ☐ Sunflower ☐ Peanut ☐ Others. Which?
3. How many days a week do you eat nuts?	☐ None ☐ 1–3 ☐ More than 3
4. How many days a week do you consume butter or margarine?	☐ None ☐ 1–3 ☐ More than 3
5. How many days a week do you consume pastries?	☐ None ☐ 1–3 ☐ More than 3
6. How many days a week do you consume sausages?	☐ None ☐ 1–3 ☐ More than 3
7. Do you take supplements rich in vitamin E, α-tocopherol (Auxina®), or a multivitamin complex (Supradyn®, Pharmaton®, Multicentrum®...)?	☐ Yes ☐ No
8. Do you take Danacol® or some product to lower cholesterol?	☐ Yes ☐ No
9. Do you take a cholesterol-lowering drug?	☐ Yes ☐ No

References

1. Bhattacharyya, A.K.; Connor, W. Beta-Sitosterolemia and xanthomatosis. *J. Clin. Investig.* **1974**, *53*, 1033–1043. [CrossRef] [PubMed]
2. Lu, K.; Lee, M.-H.; Hazard, S.; Brooks-Wilson, A.; Hidaka, H.; Kojima, H.; Ose, L.; Stalenhoef, A.F.; Mietinnen, T.; Bjorkhem, I.; et al. Two Genes That Map to the STSL Locus Cause Sitosterolemia: Genomic Structure and Spectrum of Mutations Involving Sterolin-1 and Sterolin-2, Encoded by ABCG5 and ABCG8, Respectively. *Am. J. Hum. Genet.* **2001**, *69*, 278–290. [CrossRef] [PubMed]
3. Lee, M.-H.; Lu, K.; Hazard, S.; Yu, H.; Shulenin, S.; Hidaka, H.; Kojima, H.; Allikmets, R.; Sakuma, N.; Pegoraro, R.; et al. Identification of a gene, ABCG5, important in the regulation of dietary cholesterol absorption. *Nat. Genet.* **2001**, *27*, 79–83. [CrossRef] [PubMed]
4. Patel, S.B.; Salen, G.; Hidaka, H.; Kwiterovich, P.O.; Stalenhoef, A.F.; Miettinen, T.A.; Grundy, S.M.; Lee, M.H.; Rubenstein, J.S.; Polymeropoulos, M.H.; et al. Mapping a gene involved in regulating dietary cholesterol absorption. The sitosterolemia locus is found at chromosome 2p21. *J. Clin. Investig.* **1998**, *102*, 1041–1044. [CrossRef] [PubMed]
5. Berge, K.E.; Tian, H.; Graf, G.A.; Yu, L.; Grishin, N.V.; Schultz, J.; Kwiterovich, P.; Shan, B.; Barnes, R.; Hobbs, H.H. Accumulation of dietary cholesterol in sitosterolemia caused by mutations in adjacent ABC transporters. *Science* **2000**, *290*, 1771–1775. [CrossRef] [PubMed]
6. Wang, H.H.; Liu, M.; Portincasa, P.; Wang, D.Q.H. Recent advances in the critical role of the sterol efflux transporters ABCG5/G8 in health and disease. *Adv. Exp. Med. Biol.* **2020**, *1276*, 105–136. [PubMed]
7. Teupser, D.; Baber, R.; Ceglarek, U.; Scholz, M.; Illig, T.; Gieger, C.; Holdt, L.M.; Leichtle, A.; Greiser, K.H.; Huster, D.; et al. Genetic Regulation of Serum Phytosterol Levels and Risk of Coronary Artery Disease. *Circ. Cardiovasc. Genet.* **2010**, *3*, 331–339. [CrossRef] [PubMed]
8. Berge, K.E.; von Bergmann, K.; Lutjohann, D.; Guerra, R.; Grundy, S.M.; Hobbs, H.H.; Cohen, J.C. Heritability of plasma noncholesterol sterols and relationship to DNA sequence polymorphism in ABCG5 and ABCG8. *J. Lipid Res.* **2002**, *43*, 486–494. [CrossRef] [PubMed]
9. Plat, J.; Bragt, M.C.; Mensink, R.P. Common sequence variations in ABCG8 are related to plant sterol metabolism in healthy volunteers. *J. Lipid Res.* **2005**, *46*, 68–75. [CrossRef]
10. Gylling, H.; Hallikainen, M.; Pihlajamäki, J.; Ågren, J.; Laakso, M.; Rajaratnam, R.A.; Rauramaa, R.; Miettinen, T.A. Polymorphisms in the ABCG5 and ABCG8 genes associate with cholesterol absorption and insulin sensitivity. *J. Lipid Res.* **2004**, *45*, 1660–1665. [CrossRef]
11. gNomad. Available online: https://ldlink.nih.gov/%E2%80%A6ldpair (accessed on 12 June 2023).
12. Helgadottir, A.; Thorleifsson, G.; Alexandersson, K.F.; Tragante, V.; Thorsteinsdottir, M.; Eiriksson, F.F.; Gretarsdottir, S.; Björnsson, E.; Magnusson, O.; Sveinbjornsson, G.; et al. Genetic variability in the absorption of dietary sterols affects the risk of coronary artery disease. *Eur. Heart J.* **2020**, *41*, 2618–2628. [CrossRef] [PubMed]
13. Kurvinen, A.; Nissinen, M.J.; Gylling, H.; Miettinen, T.A.; Lampela, H.; Koivusalo, A.I.; Rintala, R.J.; Pakarinen, M.P. Effects of long-term parenteral nutrition on serum lipids, plant sterols, cholesterol metabolism, and liver histology in pediatric intestinal failure. *J. Pediatr. Gastroenterol. Nutr.* **2011**, *53*, 440–446. [CrossRef] [PubMed]
14. Pianese, P.; Salvia, G.; Campanozzi, A.; D'Apolito, O.; Russo, A.D.; Pettoello-Mantovani, M.; Corso, G. Sterol profiling in red blood cell membranes and plasma of newborns receiving total parenteral nutrition. *J. Pediatr. Gastroenterol. Nutr.* **2008**, *47*, 645–651. [CrossRef] [PubMed]
15. Clayton, P.T.; Bowron, A.; Mills, K.A.; Massoud, A.; Casteels, M.; Milla, P.J. Phytosterolemia in children with parenteral nutrition—Associated cholestatic liver disease. *Gastroenterology* **1993**, *105*, 1806–1813. [CrossRef] [PubMed]
16. Kurvinen, A.; Nissinen, M.J.; Andersson, S.; Korhonen, P.; Ruuska, T.; Taimisto, M.; Kalliomäki, M.; Lehtonen, L.; Sankilampi, U.; Arikoski, P.; et al. Parenteral plant sterols and intestinal failure–associated liver disease in neonates. *J. Pediatr. Gastroenterol. Nutr.* **2012**, *54*, 803–811. [CrossRef] [PubMed]
17. Llop-Talaveron, J.; Badía-Tahull, M.; Lozano-Andreu, T.; Rigo-Bonnin, R.; Virgili-Casas, N.; Farran-Teixidó, L.; Miró-Martín, M.; Suárez-Lledó, A.; Leiva-Badosa, E. Phytosterolemia and γ-glutamyl transferase in adults with parenteral nutrition: Fish versus vegetal lipids: A randomized clinical trial. *Nutrition* **2019**, *70*, 110587. [CrossRef] [PubMed]
18. Llop-Talaveron, J.; Leiva-Badosa, E.; Alia-Ramos, P.; Rigo-Bonnin, R.; Virgili-Casas, N.; Farran-Teixidor, L.; Miró-Martín, M.; Garrido-Sanchez, L.; Suárez-Lledó, A.; Badía-Tahull, M.B. Genetic factors associated with alterations in liver function test results in adult hospitalized patients treated with parenteral nutrition: A substudy of a clinical trial. *Nutrition* **2022**, *93*, 111507. [CrossRef] [PubMed]
19. Merino, J.; Masana, L.; Guijarro, C.; Ascaso, J.; Lagares, M.; Civeira, F. Recomendaciones para la utilización clínica de los alimentos enriquecidos con fitosteroles/fitostanoles en el manejo de la hipercolesterolemia. *Clin. Investig. Arterioscler.* **2014**, *26*, 147–158. [PubMed]
20. Schroor, M.M.; Mokhtar, F.B.A.; Plat, J.; Mensink, R.P. Associations between SNPs in intestinal cholesterol absorption and endogenous cholesterol synthesis genes with cholesterol metabolism. *Biomedicines* **2021**, *9*, 1475. [CrossRef]
21. Silbernagel, G.; Chapman, M.J.; Genser, B.; Kleber, M.E.; Fauler, G.; Scharnagl, H.; Grammer, T.B.; Boehm, B.O.; Mäkelä, K.-M.; Kähönen, M.; et al. High intestinal cholesterol absorption is associated with cardiovascular disease and risk alleles in ABCG8 and ABO: Evidence from the LURIC and YFS cohorts and from a meta-analysis. *J. Am. Coll. Cardiol.* **2013**, *62*, 291–299. [CrossRef]

22. Jakulj, L.; Vissers, M.N.; Tanck, M.W.; Hutten, B.A.; Stellaard, F.; Kastelein, J.J.; Dallinga-Thie, G.M. ABCG5/G8 polymorphisms and markers of cholesterol metabolism: Systematic review and meta-analysis. *J. Lipid Res.* **2010**, *51*, 3016–3023. [CrossRef] [PubMed]
23. Assmann, G.; Cullen, P.; Erbey, J.; Ramey, D.R.; Kannenberg, F.; Schulte, H. Plasma sitosterol elevations are associated with an increased incidence of coronary events in men: Results of a nested case-control analysis of the Prospective Cardiovascular Münster (PROCAM) study. *Nutr. Metab. Cardiovasc. Dis.* **2006**, *16*, 13–21. [CrossRef] [PubMed]
24. Windler, E.; Zyriax, B.-C.; Kuipers, F.; Linseisen, J.; Boeing, H. Association of plasma phytosterol concentrations with incident coronary heart disease. Data from the CORA study, a case-control study of coronary artery disease in women. *Atherosclerosis* **2009**, *203*, 284–290. [CrossRef]
25. Pinedo, S.; Vissers, M.N.; von Bergmann, K.; Elharchaoui, K.; Lütjohann, D.; Luben, R.; Wareham, N.J.; Kastelein, J.J.P.; Khaw, K.-T.; Boekholdt, S.M. Plasma levels of plant sterols and the risk of coronary artery disease: The prospective EPIC-Norfolk Population Study. *J. Lipid Res.* **2007**, *48*, 139–144. [CrossRef] [PubMed]
26. Weingärtner, O.; Bogeski, I.; Kummerow, C.; Schirmer, S.H.; Husche, C.; Vanmierlo, T.; Wagenpfeil, G.; Hoth, M.; Böhm, M.; Lütjohann, D.; et al. Plant sterol ester diet supplementation increases serum plant sterols and markers of cholesterol synthesis, but has no effect on total cholesterol levels. *J. Steroid Biochem. Mol. Biol.* **2017**, *169*, 219–225. [CrossRef]
27. Fassbender, K.; Lütjohann, D.; Dik, M.G.; Bremmer, M.; König, J.; Walter, S.; Liu, Y.; Letièmbre, M.; von Bergmann, K.; Jonker, C. Moderately elevated plant sterol levels are associated with reduced cardiovascular risk—The LASA study. *Atherosclerosis* **2008**, *196*, 283–288. [CrossRef]
28. Silbernagel, G.; Fauler, G.; Renner, W.; Landl, E.M.; Hoffmann, M.M.; Winkelmann, B.R.; Boehm, B.O.; März, W. The relationships of cholesterol metabolism and plasma plant sterols with the severity of coronary artery disease. *J. Lipid Res.* **2009**, *50*, 334–341. [CrossRef] [PubMed]
29. Ellegård, L.; Sunesson, Å.; Bosaeus, I. High serum phytosterol levels in short bowel patients on parenteral nutrition support. *Clin. Nutr.* **2005**, *24*, 415–420. [CrossRef] [PubMed]
30. Llop, J.M.; Virgili, N.; Moreno-Villares, J.M.; García-Peris, P.; Serrano, T.; Forga, M.; Solanich, J.; Pita, A.M. Phytosterolemia in parenteral nutrition patients: Implications for liver disease development. *Nutrition* **2008**, *24*, 1145–1152. [CrossRef]
31. Gylling, H.; Puska, P.; Vartiainen, E.; Miettinen, T. Retinol, vitamin D, carotenes and α-tocopherol in serum of a moderately hypercholesterolemic population consuming sitostanol ester margarine. *Atherosclerosis* **1999**, *145*, 279–285. [CrossRef]
32. Chalasani, N.; Younossi, Z.; LaVine, J.E.; Diehl, A.M.; Brunt, E.M.; Cusi, K.; Charlton, M.; Sanyal, A.J. The diagnosis and management of non-alcoholic fatty liver disease: Practice Guideline by the American Association for the Study of Liver Diseases, American College of Gastroenterology, and the American Gastroenterological Association. *Hepatology* **2012**, *55*, 2005–2023. [CrossRef]
33. Wanten, G.; Beunk, J.; Naber, A.; Swinkels, D. Tocopherol isoforms in parenteral lipid emulsions and neutrophil activation. *Clin. Nutr.* **2002**, *21*, 417–422. [CrossRef]
34. Pironi, L.; Guidetti, M.; Zolezzi, C.; Fasano, M.C.; Paganelli, F.; Merli, C.; Bersani, G.; Pizzoferrato, A.; Miglioli, M. Peroxidation potential of lipid emulsions after compounding in all-in-one solutions. *Nutrition* **2003**, *19*, 784–788. [CrossRef]
35. Miettinen, T.; Hallikainen, M.; Raitakari, O.; Viikari, J.; Gylling, H. Twenty-one year tracking of serum non-cholesterol sterols. The Cardiovascular Risk in Young Finns study. *Nutr. Metab. Cardiovasc. Dis.* **2009**, *19*, 525–531. [CrossRef]
36. Racette, S.B.; Lin, X.; Ma, L.; Ostlund, J.R.E. Natural Dietary Phytosterols. *J. AOAC Int.* **2015**, *98*, 679–684. [CrossRef] [PubMed]
37. Escurriol, V.; Cofán, M.; Moreno-Iribas, C.; Larrañaga, N.; Martínez, C.; Navarro, C.; Rodríguez, L.; González, C.A.; Corella, D.; Ros, E. Phytosterol plasma concentrations and coronary heart disease in the prospective Spanish EPIC cohort. *J. Lipid Res.* **2010**, *51*, 618–624. [CrossRef] [PubMed]
38. Ras, R.T.; Geleijnse, J.M.; Trautwein, E.A. LDL-cholesterol-lowering effect of plant sterols and stanols across different dose ranges: A meta-analysis of randomised controlled studies. *Br. J. Nutr.* **2014**, *112*, 214–219. [CrossRef]
39. Andersson, S.W.; Skinner, J.; Ellegård, L.; Welch, A.A.; Bingham, S.; Mulligan, A.; Andersson, H.; Shaw, K.-T. Intake of dietary plant sterols is inversely related to serum cholesterol concentration in men and women in the EPIC Norfolk population: A cross-sectional study. *Eur. J. Clin. Nutr.* **2004**, *58*, 1378–1385. [CrossRef] [PubMed]
40. Klingberg, S.; Ellegård, L.; Johansson, I.; Hallmans, G.; Weinehall, L.; Andersson, H.; Winkvist, A. Inverse relation between dietary intake of naturally occurring plant sterols and serum cholesterol in northern Sweden. *Am. J. Clin. Nutr.* **2008**, *87*, 993–1001. [CrossRef]
41. Gylling, H.; Plat, J.; Turley, S.; Ginsberg, H.N.; Ellegård, L.; Jessup, W.; Jones, P.J.H.; Lütjohann, D.; Maerz, W.; Masana, L.; et al. Plant sterols and plant stanols in the management of dyslipidaemia and prevention of cardiovascular disease. *Atherosclerosis* **2014**, *232*, 346–360. [CrossRef]
42. Heinemann, T.; Axtmann, G.; Von Bergmann, K. Comparison of intestinal absorption of cholesterol with different plant sterols in man. *Eur. J. Clin. Investig.* **1993**, *23*, 827–831. [CrossRef] [PubMed]
43. Yoshida, H.; Tada, H.; Ito, K.; Kishimoto, Y.; Yanai, H.; Okamura, T.; Ikewaki, K.; Inagaki, K.; Shoji, T.; Bujo, H.; et al. Reference intervals of serum non-cholesterol sterols by gender in healthy Japanese individuals. *J. Atheroscler. Thromb.* **2020**, *27*, 409–417. [CrossRef] [PubMed]
44. Wu, A.H.; Ruan, W.; Todd, J.; Lynch, K.L. Biological variation of β-sitosterol, campesterol, and lathosterol as cholesterol absorption and synthesis biomarkers. *Clin. Chim. Acta* **2014**, *430*, 43–47. [CrossRef] [PubMed]

45. Salen, G.; Ahrens, E.H.; Grundy, S.M. Metabolism of beta-sitosterol in man. *J. Clin. Investig.* **1970**, *49*, 952–967. [CrossRef]
46. Miettinen, T.A.; Tilvis, R.S.; Kesäniemi, Y.A. Serum plant sterols and cholesterol precursors reflect cholesterol absorption and synthesis in volunteers of a randomly selected male population. *Am. J. Epidemiol.* **1990**, *131*, 20–31. [CrossRef]

nutrients

MDPI

Article

Genetic Variants in *CD36* Involved in Fat Taste Perception: Association with Anthropometric and Clinical Parameters in Overweight and Obese Subjects Affected by Type 2 Diabetes or Dysglycemia—A Pilot Study

Marica Franzago [1,2], Paola Borrelli [3], Marta Di Nicola [3], Liborio Stuppia [2,4] and Ester Vitacolonna [1,2,*]

1 Department of Medicine and Aging, School of Medicine and Health Sciences, "G. D'Annunzio" University, Via dei Vestini, Chieti-Pescara, 66100 Chieti, Italy; marica.franzago@unich.it

2 Center for Advanced Studies and Technology (CAST), "G. D'Annunzio" University, Chieti-Pescara, 66100 Chieti, Italy; stuppia@unich.it

3 Laboratory of Biostatistics, Department of Medical, Oral and Biotechnological Sciences, "G. D'Annunzio" University, Chieti-Pescara, 66100 Chieti, Italy; paola.borrelli@unich.it (P.B.); marta.dinicola@unich.it (M.D.N.)

4 Department of Psychological, Health and Territorial Sciences, School of Medicine and Health Sciences, "G. D'Annunzio" University, Chieti-Pescara, 66100 Chieti, Italy

* Correspondence: e.vitacolonna@unich.it

Abstract: Obesity and overweight represent a growing health problem worldwide. Genes regulating the intake and metabolism of different nutrients can positively or negatively influence the efficacy of nutritional interventions against obesity and its complications. The aim of this study was to assess changes in anthropometric and clinical parameters and the adherence to a Mediterranean diet (MedDiet) over time in relation to nutrigenetic variants in overweight or obese subjects affected by Type 2 Diabetes (T2D) or dysglycemia, who were included in a nutritional program. A total of 23 subjects were included in this study. Clinical parameters, physical activity levels, and the adherence to a MedDiet were evaluated at baseline, at 6 (T6), and at 12 months (T12) during and after a diet/lifestyle intervention. In a single blood sample from each subject, rs1984112 (*A>G*) and rs1761667 (*G>A*) in *CD36*; rs7950226 (*G>A*) in *BMAL1*; and rs1801260 (*A*>G), rs4864548 (*A>G*), and rs3736544 (*G>A*) in *CLOCK* were genotyped with Real-Time PCR. Significant associations were observed between *CD36* rs1761667 and weight ($p = 0.025$), hip circumference ($p = 0.042$), triglycerides ($p = 0.047$), and HbA1c ($p = 0.012$) at baseline. Moreover, the genotype *AA* in *CD36* rs1761667 was significantly associated with a lower BMI when compared to *G* carriers at baseline, at T6, and also at T12. In addition, subjects with the *AA* genotype at *CD36* rs1984112 had significantly lower levels of HbA1c ($p = 0.027$) than the *GG* and *AG* genotypes at baseline. These results show that variants in *CD36* can have an impact on anthropometric and clinical parameters in overweight or obese subjects affected by T2D or dysglycemia, and that it might influence the success of the diet/lifestyle intervention.

Keywords: *CD36*; *CLOCK*; *BMAL1*; rs1761667; rs1984112; gene–diet interaction; nutrition; type 2 diabetes; obesity; nutrigenetics

1. Introduction

Overweight and obesity, which are chronic, progressive, and relapsing conditions associated with an elevated risk of non-communicable diseases (NCDs, such as cardiovascular disease, type 2 diabetes (T2D), hypertension, and metabolic syndrome), have become a major global public health challenge [1,2]. Environmental and social factors, as well as the genetic susceptibility represented by the individual's genotype, can influence an individual's predisposition to the development and maintenance of obesity [1].

The increased global burden of obesity requires specific strategies to prevent weight gain, to induce weight loss, and to improve obesity comorbidities. Unfortunately, successful weight loss with behavioral and nutritional interventions, as well as long-term weight maintenance, are difficult to achieve. In this context, the effects of several single nucleotide polymorphisms (SNPs) located near or within genes regulating food intake, lipid metabolism, glucose homeostasis, and insulin signaling on metabolic improvement, weight gain/loss, and insulin resistance have been demonstrated [3,4]. Interestingly, disturbances in the circadian gene network can lead to the onset and development of obesity and some accompanying comorbidities [5]. In fact, SNPs located in circadian-related genes such as Circadian Locomotor Output Cycles Kaput (*CLOCK*) and Brain and Muscle Aryl Hydrocarbon Receptor Nuclear Translocator-Like Protein-1 (*BMAL1*), which interacts with both dietary intake and obesogenic behavior, can affect metabolic health [6–8], modifying body weight regulation [9,10]. In this regard, from a nutrigenetics point of view, minor allele *C* carriers of rs1801260 in the *CLOCK* gene have lower body weight loss than *TT* carriers [11].

In addition, it is known that obese individuals can display fat chemosensory dysfunction, and that the SNPs correlated with fat chemosensation are located in the *CD36* gene [12].

The *CD36* gene, which codes for an integral membrane glycoprotein identified as a taste receptor for fat [13], is intimately involved in several processes related to fatty acid and lipid metabolism-sensing in the organism. Genetic variants of this gene can contribute to an increased risk of obesity by modifying an individual's food preference and intake [14–17]. Among these, rs1761667, which is characterized by a substitution of allele *G* for *A*, has been associated with a decrease in sensitivity to fatty acids [15,18–20], a decrease in the expression of CD36 protein, and a decrease in metabolism [21–23]. A recent study showed that adherence to a healthy dietary pattern (a diet with high fiber, fish, and dairy products) can affect cardiometabolic risk factors and MetS risk in the *A*-allele carrier in rs1761667 [24].

Thus, gene–diet interactions seem to play an important role in the treatment of obesity, but are typically only partially assessed nowadays. Thus, the aim of this study is to evaluate the effects of a nutritional and lifestyle intervention based on nutrigenetic variants in candidate genes (namely, *CD36* rs1984112 *A>G*, *CD36* rs1761667 *G>A*, *BMAL1* rs7950226 *G>A*, *CLOCK* rs1801260 *A>G*, *CLOCK* rs4864548 *A>G*, and *CLOCK* rs3736544 *G>A*) on the anthropometric and clinical parameters of 23 overweight or obese subjects affected by T2D or impaired glucose regulation (IGR) over a one-year period. Variants located in the genes involved in clock systems (*CLOCK* and *BMAL1*) were selected, considering that circadian disruptions may contribute to different metabolic-related traits and that polymorphisms in the *CD36* gene that are related to lipid detection may be associated with interindividual variability in body weight regulation. To the best of our knowledge, this is the first study on overweight or obese subjects affected by T2D or IGR that is based on this nutrigenetic panel, evaluating not only the success of body weight loss but also the feasibility of a personalized nutritional approach.

2. Materials and Methods

2.1. Study Design and Participants

A total of 23 overweight or obese individuals affected by T2D or IGR were recruited at the Diabetes, Nutrition, and Metabolism Unit at the "Gabriele d'Annunzio" University Hospital in Chieti, Italy. This study received the approval of the Ethics Committee of the Province di Chieti and Pescara, in accordance with the Helsinki Declaration. Before undertaking the protocol, all objectives and modalities were clarified to the participants, and written informed consent was therefore obtained.

The inclusion criteria were overweight or obese (BMI $\geq$ 25 and $\geq$ 30 kg/m^2, respectively) subjects (male and female, adults: age $\geq$ 18) affected by T2D or Impaired Glucose Regulation (IGR; Impaired Fasting Glucose or Impaired Glucose Tolerance).

The exclusion criteria were as follows: subjects suffering from Type 1 Diabetes, Eating Behavior Disorders, impaired renal and hepatic function, or other conditions that might have interfered with the development and completion of the protocol.

2.2. Educational and Nutritional Intervention

At baseline (T0), all subjects participated in an educational and nutritional program, in which each individual was actively involved in face-to-face individual and group-based interventions to promote a healthy diet and lifestyle.

In detail, the first visit consisted of a clinical check-up and the collection of the individual's medical history, which included data on demographic characteristics. In addition, each participant was instructed to complete a self-monitoring diary [25–28]. The diaries were reviewed by a physician during the next meeting. In addition to the face-to-face individual intervention, educational group sessions with small groups of up to 10 people were conducted in the presence of the physician. The group-based intervention was crucial to involve the participants in improving their lifestyle. This educational and nutritional program included three main goals to achieve, as previously described [29]: (i) improve the composition of meals, prioritising vegetables as well as whole foods and limiting highly processed foods; (ii) weight loss of at least 7–10%; (iii) plan at least 150 minutes a week of moderate intensity physical activity [25–28,30].

2.3. Anthropometric and Clinical Measurements

Data on anthropometric (body weight and height, BMI, waist and hip circumferences) and clinical parameters (including fasting glucose, HbA1c (glycated hemoglobin A1c), lipid profile, and blood pressure) were collected at baseline (T0), at 6 months (T6) and at 12 months (T12) after the nutritional education program.

2.4. Questionnaires

At baseline, circadian typology (Chronotype)—defined as the interindividual circadian attitude—was assessed using a 5-item version of the Morningness–Eveningness Questionnaire (MEQr). According to Italian cut-off criteria, the MEQr generates three categories of chronotypes, usually divided by the terms: (i) "morning" (19–25 score), (ii) "evening" (4–10 score), and (iii) "intermediate" types (11–18 score) [31]. The MEQr characterizes subjects based on individual differences in wake/sleep cycle patterns and the time of the day that people feel or perform best.

Adherence to the Mediterranean diet (MedDiet) was assessed using a validated 14-item questionnaire (PREDIMED) that provides a range of possible scores, specifically: no adherence (score ≤ 5), medium adherence ($6 \leq$ score ≤ 9), and maximum adherence (score ≥ 10) [32]. Moreover, levels of physical activity were evaluated using a short version of the International Physical Activity Questionnaire (IPAQ), which showed three different intensity levels, namely: low, moderate, and high PA [33]. Physical activity and MedDiet adherence were evaluated at baseline, T6, and T12.

2.5. Gene and SNP Selection

Nutrigenetic variants from three genes, identified in previous studies as associated with T2D, obesity, lipid metabolism, and dietary intake were selected. Specifically, two of these variants in the *CD36* gene, namely rs1984112 (*A*>*G*) and rs1761667 (*G*>*A*), were involved in fat intake regulation [34,35]. Moreover, rs7950226 (G>A) in *BMAL1*, rs1801260 (*A*>*G*), rs4864548 (*A*>*G*), and rs3736544 (*G*>*A*) in *CLOCK* were involved in obesity, CVD, MetS, sleep reduction, alterations in eating behaviors, and evening preference [9,36–43].

Genetic Analysis

Blood from each participant was collected for genotype analyses in a tube containing EDTA and was stored at +4 °C before analysis.

The genetic analysis was conducted at the Laboratory of Molecular Genetics, School of Medicine and Health Sciences, "G. d'Annunzio" University of Chieti-Pescara. Genomic DNA was extracted from peripheral blood lymphocytes using a MagPurix 12sAutomatedNucleicAcid Purification System (Zinexts Life Science Corp., New Taipei City, Taiwan). Nucleic acids were quantified using the Qubit assay kit on a Qubit 4 Fluorometer (Invitrogen, Thermo Fisher, Waltham, MA, USA).

To genotype the polymorphisms, molecular analyses were performed using a TaqMan SNP Genotyping Assay according to the manufacturer's instructions (ThermoFisher Scientific, Foster City, CA, USA). The details of TaqMan SNP Genotyping assays are available in Table 1.

Table 1. The TaqMan SNP Genotyping Assays.

Genetic Variants	Location	Assay
CD36 rs1984112 (*A>G*)	Chr.7: 80613604 on GRCh38	C__12093946_10
CD36 rs1761667 (*G>A*)	Chr.7: 80615623 on GRCh38	C___8314999_10
CLOCK rs1801260 (*A>G*)	Chr.4: 55435202 on GRCh38	C___8746719_20
CLOCK rs4864548 (*A>G*)	Chr.4: 55547636 on GRCh38	C__11821276_10
BMAL1 rs7950226 (*G>A*)	Chr.11: 13296592 on GRCh38	C__11578388_10
CLOCK rs3736544 (*G>A*)	Chr.4: 55443825 on GRCh38	C__22273263_10

In detail, approximately 20 ng of DNA, 0.125 μL of TaqMan 40× concentration assay, and 2.5 μL of TaqMan Mastermix were used to amplify DNA sequences of interest on the QuantStudio 5 Real-Time PCR System (Applied Biosystem, Foster City, CA, USA). PCR conditions were 95 °C for 10 min, 40 cycles of 95 °C for 15 s, and 60 °C for 1 min.

2.6. Statistical Analysis

Descriptive analysis was carried out using the median and interquartile range (IQR) for quantitative variables, while frequencies and percentages were used to describe the qualitative variables. Normality distribution was tested using the Shapiro–Wilk test. To evaluate the relationships between qualitative variables, a Pearson chi square test and/or Fisher's test was assessed. However, for quantitative variables, the Friedman test was used to determine the differences between the medians of the three time periods (Baseline, T6, and T12) and the Wilcoxon rank-sum (Mann–Whitney) test or Kruskal–Wallis test was performed to assess the differences between nutritional parameters at baseline and in genotypes (for additive, dominant, and recessive models). For significant trends, this analysis was followed by the Sign test or Dunn's test for comparisons between median pairs for the identification of significant differences. The Bonferroni's correction for multiple comparisons tests was applied. Several linear mixed models were used to determine the differences between male and female patients, between time periods (baseline, T6, and T12), and their interactions with the anthropometric and clinical data. Several linear mixed models were also used to determine the differences between different genotypes, between different time periods (Baseline, T6, and T12), and their interaction with the following variables: weight, BMI, and PREDIMED.

A statistical significance was set at the level of ≤0.05, unless adjustment for multiple comparisons was needed. All analyses were performed using Stata software v17 (StataCorp, College Station, TX, USA).

3. Results

The Demographic and clinical characteristics of the cohort of participants at baseline, at T6, and at T12 of the nutritional intervention are reported in Tables 2 and 3, respectively.

Table 2. Sociodemographic characteristics of participants.

Variable	Baseline
Age (yr)	65.0 (57.0–66.0)
Gender	
- Female	12 (52.2%)
- Male	11 (47.8%)
Employment	
- Employed	22 (95.7%)
- Unemployed	1 (4.3%)

n (%) or median and interquartile range (IQR) are shown when appropriate.

Table 3. Anthropometric and clinical data of participants at baseline, T6, and T12.

Variable	Baseline	T6	T12	*p*-Value [a]
Weight (Kg)	93.0 (78.0–103.0)	92.0 (79.0–101.0) *	91.5 (79.5–104.0) *	**0.002**
BMI (kg/m^2)	31.8 (28.1–37.8)	33.7 (27.9–37.2) *	30.4 (26.1–36.1) *	**0.002**
Waist circumference (cm)	108.5 (100.0–118.0)	107.0 (97.0–113.0)	109.0 (93.0–116.0)	0.152
Hip circumference (cm)	113.0 (104.0–126.0)	114.5 (101.5–123.0)	112.0 (99.0–120.0)	0.452
WHR	1.0 (0.9–1.0)	1.1 (1.0–1.1)	0.9 (0.9–1.0)	0.178
Systolic blood pressure (mmHg)	130.0 (110.0–150.0)	130.0 (120.0–140.0)	125.0 (120.0–137.5)	0.717
Diastolic blood pressure (mmHg)	80.0 (70.0–90.0)	80.0 (75.0–85.0)	80.0 (70.0–80.0)	0.494
PREDIMED	7.0 (7.0–8.0)	9.0 (9.0–10.0) *	8.5 (8.0–10.0) *	**0.002**
PREDIMED CLASS				0.301
- No adherence	5.3%	0.0%	0.0%	
- Adherence	78.9%	58.8%	72.2%	
- Max adherence	15.8%	41.2%	27.8%	
IPAQ				0.318
- Low	52.6%	29.4%	22.2%	
- Moderate	31.6%	47.1%	38.9%	
- High	15.8%	23.5%	38.9%	
Fasting blood glucose (mg/dL)	115.0 (105.0–140.0)	107.0 (102.0–108.0)	110.0 (102.0–125.0)	0.223
Hba1c	6.5 (5.9–7.5)	6.1 (6.0–6.3)	6.3 (5.5–6.7)	0.350
Total cholesterol (mg/dL)	199.0 (183.0–217.5)	208.5 (197.0–238.5)	193.0 (160.0–223.0)	0.751
HDL (mg/dL)	41.0 (37.0–55.0)	48.5 (41.0–55.0)	46.0 (41.0–51.0)	0.135
TG (mg/dL)	139.5 (93.5–214.5)	118.5 (94.5–187.5)	120.4 (96.0–160.7)	0.900
LDL (mg/dL)	120.4 (101.2–138.0)	132.7 (121.6–160.7)	123.8 (90.8–137.0)	0.913

n (%) or median and interquartile range (IQR) are shown when appropriate. BMI, body mass index; HDL-C, high-density lipoprotein cholesterol; LDL-C, low-density lipoprotein cholesterol; TG, triglycerides. [a] *p*-values are for Pearson's chi-square test or Friedman test; * p-value < $\alpha/3$ for Bonferroni multiple testing correction vs. Baseline. Statistically significant values are given in bold.

A total of 23 subjects (n = 23, 11 males and 12 females) were included in the study. The median age of the participants was 65.0 (IQR 57.0–66.0) years. Regarding the distribution of chronotypes, 21.7% were a "morning-type" and 78.3% were an "intermediate-type".

Our results showed significant variations over time for weight (p = 0.002) and BMI (p = 0.002; Table 3). In addition, the median PREDIMED score was 7.0 and 8.5 at baseline and at the end of the study, respectively (p = 0.002). Supplementary Table S1 shows the anthropometric and clinical data for females and males, separately, at baseline, T6, and T12.

Females had a significantly lower hip circumference (p = 0.034) and higher PREDIMED scores (p = 0.041) over time when compared to males (Supplementary Table S1).

Supplementary Figures S1–S3 summarize the distribution of genotypes based on additive, dominant, and recessive inheritance genetic models. The distribution of the tested SNP allele frequencies in the participant cohort compared to that in the European and general population reported to 1000 Genomes (dbSNP Short Genetic Variations) are shown in Supplementary Table S2.

Regarding *CD36* rs1761667 *(G>A)*, the additive genetic model (*AA* vs. *AG* vs. *GG*) showed that subjects carrying the *AA* genotype had a significantly lower hip circumference (*AA* 102.5 (IQR 101.0–104.0) vs. *AG* 123.0 (IQR 112.5–128.0) vs. *GG* 109.0 (IQR 104.0–114.0), p = 0.042; in particular, *AA* vs. *AG*, p = 0.028), weight (*AA* 82.8 (IQR 78.0–83.0) vs. *AG* 102.0 (IQR 92.8–114.3) vs. *GG* 89.5 (IQR 66.0–92.5), p = 0.025; in particular, *AA* vs. *AG*, p = 0.027), BMI (*AA* 28.1 (IQR 27.3–28.7) vs. *AG* 37.2 (IQR 32.4–39.9) vs. *GG* 31.3 (IQR 28.2–35.9), p = 0.012; in particular, *AA* vs. *AG* p = 0.006), and HbA1c (*AA* 5.8 (IQR 5.6–6.3) vs. *AG* 6.5 (IQR 5.9–7.4) vs. *GG* 7.6 (IQR 7.5–8.4), p = 0.012; in particular, *AA* vs. *GG*, p = 0.004) when compared to the *GG* and *AG* carriers at baseline.

Furthermore, using a recessive inheritance model ((*GA*+*GG*) vs. (*AA*)), the subjects carrying the *G* allele in *CD36* rs1761667 had a higher BMI (35.9 (IQR 31.3–38.0) vs. 28.1 (IQR 27.3–28.7), p = 0.009, respectively), HbA1c (7.2 (IQR 6.3–7.7) vs. 5.8 (IQR 5.6–6.3), p = 0.013, respectively) and hip circumference (114.0 (IQR 110.0–126.0) vs. 102.5 (IQR 101.0–104.0), p = 0.045, respectively) than *AA* carriers at baseline.

In addition, the dominant inheritance model ((*GA*+*AA*) vs. (*GG*)) of *CD36* rs1761667 demonstrated that carriers of the *A*-allele reported significantly lower levels of triglycerides (119.5 (IQR 90.0–189.5) vs. 253.5 (IQR 174.0–380.5), p = 0.047), HbA1c (6.3 (IQR 5.9–6.6) vs. 7.6 (IQR 7.5–8.4), p = 0.013) when compared to *GG* carriers at baseline.

However, regarding *CD36* rs1984112 (*A*>*G*), subjects with the *AA* genotype had significantly lower levels of HbA1c (*AA* 5.8 (IQR 5.6–6.3) vs. *AG* 6.6 (IQR 6.3–7.5) vs. *GG* 8.3 (IQR 7.4–9.2), p= 0.028; in particular, *AA* vs. *GG*, p = 0.027) than *GG* and *AG* genotypes at baseline.

Moreover, mixed linear models for additive and recessive inheritance models of *CD36* rs1761667 showed a significant relationship between genotype and BMI, as well as between the latter and the time trend (Tables 4 and 5). In particular, homozygous subjects for the *A* allele had a lower BMI compared to *G* carriers at baseline, at T6, and at T12 (Tables 4 and 5). We also analyzed participants' dietary habits through four items of the PREDIMED related to fat intake (use of olive oil, sofrito, consumption of red/processed meats and butter/cream/margarine). A significant fat consumption reduction after the nutritional intervention adjusted for CD36 rs1761667 genotypes (p = 0.007) was found.

For the other genetic variants, a statistically significant difference was found only in the temporal trend, excluding *CLOCK* rs3736544 in the recessive inheritance model (Tables 4 and 5).

No significant differences were detected related to SNPs in the *CLOCK* and *BMAL1* genes. Regarding questionnaires relating all the tested genetic variants, no differences were detected between PREDIMED, MEQr, or IPAQ scores assuming dominant, recessive, or additive genetic inheritance models. In addition, the interaction terms between genetic variants and time were also not significant.

Table 4. Mixed linear model for additive genetic model. Relationship between the subjects' BMI at baseline with those at T6 and T12 and its association with SNPs.

				p-Value		
CD36* rs1984112 *A>G	**AA**	**AG**	**GG**	**Genotype [a]**	**Time [b]**	**Interaction [c]**
BMI						
Baseline	28.6 (27.6–29.4)	36.2 (31.9–38.0)	29.7 (28.2–31.3)	0.124	**0.019**	0.964
T6	27.8 (27.3–29.1)	36.0 (31.6–38.0)	29.3 (28.0–30.6)			
T12	27.9 (26.1–28.2)	35.1 (30.0–36.9)	29.1 (27.8–30.4)			
CD36 rs1761667 *G>A*	AA	AG	GG			
BMI						
Baseline	28.0 (27.3–28.7)	37.2 (32.4–39.9)	31.3 (28.2–35.9)	**0.011**	**0.001**	0.968
T6	27.6 (26.5–28.4)	36.2 (32.7–39.3)	30.6 (28.0–36.0)			
T12	26.1 (25.4–27.9)	35.5 (31.9–38.3)	29.1 (25.5–32.3)			
BMAL1 rs7950226 *G>A*	AA	AG	GG			
BMI						
Baseline	34.7 (26.9–44.2)	36.5 (28.5–37.9)	30.5 (28.0–35.7)	0.567	**<0.001**	0.167
T6	40.3 (28.0–45.9)	35.9 (27.3–37.2)	30.6 (27.8–36.0)			
T12	32.3 (26.7–41.6)	35.5 (29.8–36.1)	30.0 (26.1–34.6)			
CLOCK rs1801260 *A>G*	AA	AG	GG			
BMI						
Baseline	35.9 (29.4–37.9)	29.1 (26.7–32.4)	41.1 (27.6–41.3)	0.081	**<0.001**	0.214
T6	36.0 (30.6–37.2)	28.9 (25.0–32.7)	40.3 (27.3–40.8)			
T12	34.6 (28.2–36.0)	26.7 (24.7–30.9)	38.6 (36.9–40.4)			
CLOCK rs4864548 *G>A*	AA	AG	GG			
BMI						
Baseline	37.9 (37.9–37.9)	32.1 (28.7–35.9)	30.7 (26.4–38.8)	0.753	**0.008**	0.593
T6	36.5 (36.5–36.5)	33.8 (29.1–36.0)	31.6 (27.3–40.3)			
T12	36.0 (36.0–36.0)	29.3 (27.0–34.9)	30.9 (25.6–36.9)			
CLOCK rs3736544 *G>A*	AA	AG	GG			
BMI						
Baseline	36.6 (31.9–42.0)	29.7 (28.7–35.6)	35.4 (28.4–39.9)	0.132	**0.003**	0.387
T6	37.0 (31.9–42.0)	30.6 (29.1–35.1)	35.1 (27.7–39.3)			
T12	36.1 (27.9–46.5)	29.1 (26.1–31.9)	36.4 (27.8–38.3)			

Data are expressed as median and interquartile range (IQR). Statistically significant values are given in bold. [a] Groups, for each variable, the differences have been tested between Genotypes over time. [b] Time, for each variable, the differences have been tested between baseline, T6, and T12 of the three Genotypes. [c] Probability that the effects of nutritional intervention are greater in one distinct group (interaction Time × Genotype).

Table 5. Mixed linear model for recessive inheritance model. Relationship between the subjects' BMI at baseline with those at T6 and T12 and its association with SNPs.

			p-Value		
CD36* rs1984112 *A>G	**GG**	**AA+AG**	**Genotype** [a]	**Time** [b]	**Interaction** [c]
BMI					
Baseline	29.7 (28.2–31.3)	33.0 (28.6–37.9)	0.438	0.119	0.781
T6	29.3 (28.0–30.6)	35.1 (27.8–38.0)			
T12	29.1 (27.8–30.4)	31.8 (26.1–36.1)			
CD36 rs1761667 *G>A*	AA	GG+GA			
BMI					
Baseline	28.0 (27.3–28.7)	35.9 (31.2–38.0)	**0.014**	**0.001**	0.831
T6	27.6 (26.4–28.4)	35.9 (30.6–38.0)			
T12	26.1 (25.6–27.9)	34.9 (30.0–36.9)			
BMAL1 rs7950226 *G>A*	AA	GG+GA			
BMI					
Baseline	34.8 (26.9–44.2)	31.9 (28.5–36.7)	0.400	**<0.001**	0.877
T6	40.3 (28.0–45.9)	32.7 (27.8–36.5)			
T12	32.3 (26.7–41.6)	30.4 (26.1–36.0)			
CLOCK rs1801260 *A>G*	GG	AA+AG			
BMI					
Baseline	41.1 (27.6–41.3)	31.9 (28.5–36.7)	0.362	**<0.001**	0.123
T6	40.3 (27.3–40.8)	32.7 (28.0–36.5)			
T12	38.6 (36.9–40.4)	30.2 (27.0–35.5)			
CLOCK rs4864548 *G>A*	AA	GG+GA			
BMI					
Baseline	37.9 (37.9–37.9)	31.6 (28.2–36.7)	0.462	**0.049**	0.699
T6	36.5 (36.5–36.5)	32.7 (27.9–37.6)			
T12	36.0 (36.0–36.0)	30.2 (26.1–36.1)			
CLOCK rs3736544 *G>A*	AA	GG+GA			
BMI					
Baseline	36.6 (31.9–42.0)	31.2 (28.2–37.9)	0.129	0.078	0.363
T6	37.0 (31.9–42.0)	31.6 (28.0–36.5)			
T12	36.1 (27.9–46.5)	30.2 (26.0–35.6)			

Data are expressed as median and interquartile range (IQR). Statistically significant values are given in bold. [a] Groups, for each variable, the differences have been tested between Genotypes over time. [b] Time, for each variable, the differences have been tested between baseline, T6, and T12 of the two Genotypes. [c] Probability that the effects of nutritional intervention are greater in one distinct group (interaction Time × Genotype).

4. Discussion

The main aim of this pilot study was to evaluate the effects of the interactions between some genetic variants and nutritional intervention on mid-term changes in anthropometric and clinical parameters in overweight or obese subjects affected by T2D or IGR over a one-year period.

The additive genetic model showed that *GG* and *AG* carriers at *CD36* rs1761667 presented with higher significant parameters such as BMI, hip circumference, and weight compared to subjects for homozygous the *A* allele at baseline.

CD36 is involved in several processes including oro-sensory perception of dietary lipids, inflammatory responses, angiogenesis, metabolism, and regulation of the metabolic pathways of insulin-resistance [44]. CD36 is a widely expressed glycoprotein that acts as a receptor for a several ligands, including saturated, mono-unsaturated, and poly-unsaturated fatty acids in taste bud cells, as well as being a transporter of long-chain fatty acids into adipose and muscle tissues [14,45]. In the past, the decreased expression of CD36 in the circumvallate taste buds of high-fat diet-induced obese rats has been related to diminished fatty taste sensitivity; consequently, the intake of dietary fat increases as a compensatory response [46]. Lower CD36 expression, induced by a currently unknown mechanism in the *AA* and *AG* genotypes at rs1761667, has been associated with food choices, lipid profiles, and adiposity parameters, as well as in reducing the release of peptide YY from taste bud cells [46]. These relationships may play a key role in the preference for energy-rich diets, obesity risk, and associated complications. The genotypic variation at rs1761667 has been emphasized as a reason for the disparity in fat perception between individuals [47]. On the other hand, several studies in different ethnic populations have shown ambivalent results regarding the association between rs1761667 and T2D, total fat consumption and fat taste perception, obesity, and metabolic syndrome [35,48–51].

It should be noted that there are differences in the frequency of minor alleles in different ethnic groups; in fact, *A* is the minor allele in Africans and Asians, while *G* is the minor allele in Caucasians and Americans. This may in part explain the differences in the genotype–phenotype relationship reported across studies, as rs1761667 may have ethnic-specific effects on the perception of fat-containing foods [20].

Interestingly, Pioltine et al. [52] reported that the rs1761667 in *CD36* is not related to obesity, although the *A* allele is associated with decreased fat and carbohydrate (CHO) intake in obese children and adolescents. The association of *CD36* rs1761667 with BMI and hypertension has also been studied in a cohort of adults, showing a correlation between the *AA* genotype and lower BMI as compared to *AG* and *GG* [53]. In addition, it has been shown that the *AA* genotype has a significantly lower degree of dyslipidemia, systolic blood pressure, and WC compared to *G* carriers [51]. Boghdady et al. [49] also demonstrated that the *AG* genotype is associated with coronary artery disease, raised BMI, metabolic syndrome, and T2D.

Furthermore, it has been shown that the *CD36* SNP is linked to cardiovascular risk factors, suggesting an important role in LDL-C and HDL-C metabolism [20,54–56].

In our study, *CD36* with the *AA* genotype at rs1761667 showed lower levels of TG when compared to *GG* and *AG*. This finding conflicts with the study of Karthi et al. [47], but it is consistent with a recent systematic review and meta-analysis [57]. Therefore, it is necessary to clarify the effects of taste perception and lipid transport on TG levels in subjects with *CD36* rs1761667. Nevertheless, further studies are required to understand if the modulation in TG levels observed in the *CD36* variant occurs at the level of taste perception or metabolism.

The present study demonstrated that homozygous subjects with the *A* allele at both *CD36* rs1761667 and rs1984112 had significantly lower HbA1c concentrations than those with the *GG* or *AG* genotypes at baseline. This observation corroborates the findings of previous studies [19,58], in which it has been shown that the *GG* genotype is significantly associated with higher plasma HbA1c as compared to the *AA* genotype of *CD36* rs1761667. Considering the role of CD36 in some aspects of fatty acid and lipid metabolism, it has been suggested that variants in this gene can influence energy homeostasis [53].

Considering the significant role of CD36 as a receptor defining the preference for fat, taste dysfunction may be responsible for abnormalities in food intake (leading to obesity) and may not be reversed by weight loss (predisposing to a relapse) [13].

In this view, this pilot study provides valuable insight into the role of CD36, showing that the *AA* genotype at rs1761667 has lower adiposity parameters such as BMI, hip circumference, and weight. Our results suggest that *CD36* variants may promote a protective metabolic profile, probably reducing protein expression.

This field of taste research provides a promising path to understand how genetic variants in fat taste preference contribute to eating habits as well as to health and disease. Although the specificity and mechanism of this receptor's function must still be better clarified, it is plausible to suggest that knowledge of the *CD36* variants of an individual—which preferentially favor the intake of some nutrients and adversely affect the consumption of others—may help to prevent chronic diseases and also improve the personalization of interventions against obesity and obesity-related complications. This study has limitations—in particular, the sample size. The evidence of interactions between genetic variants and the nutritional intervention estimated may be unduly small (i.e., a false negative result), and may preclude the opportunity to further examine its latent efficacy.

However, this is a pilot study, which is a small-scale version of a subsequent investigation designed to test various parameters, including the feasibility of the study protocol, as well as measures and procedures [59,60]. In this view, the capabilities of the study design, the procedures, and strategies could be taken into consideration in a larger study.

Finally, future investigations with a large population (including obese subjects without T2D as a control group) and in different ethnic groups must be conducted to confirm these findings, and to shed light on the functional role of CD36 in obesity and on the success of diet/lifestyle interventions.

Supplementary Materials: The following supporting information can be downloaded at: https://www.mdpi.com/article/10.3390/nu15214656/s1, Table S1. Anthropometric and clinical data of females and males at baseline, at T6, and T12. Table S2. Allele frequencies in subjects compared to the general and European population. Figure S1. The distribution of genotypes in each SNP based on an additive model. Figure S2. The distribution of genotypes in each SNP based on a dominant model. Figure S3. The distribution of genotypes in each SNP based on a recessive model.

Author Contributions: M.F.: conceptualization; methodology; formal analysis; investigation; data curation; writing—original draft; review and editing. P.B.: methodology; software; formal analysis; data curation. M.D.N.: methodology; software; formal analysis; validation; editing. L.S.: review and editing; supervision. E.V.: conceptualization; methodology; recruitment of patients and clinical evaluation and treatment; data curation; validation; review and editing; supervision; project administration; funding acquisition. All authors have read and agreed to the published version of the manuscript.

Funding: This research was supported by the Italian Ministry of Health (Ministero della Salute, MOOD Project- Grant D73C22001670006).

Institutional Review Board Statement: This study was conducted in accordance with the Declaration of Helsinki and was approved by the Ethics Committee of the Province di Chieti and Pescara (protocol code: richxrked and date of approval: 12 December 2019).

Informed Consent Statement: Informed consent was obtained from all subjects involved in the study.

Data Availability Statement: The data that support the findings of this study are available from the corresponding author upon reasonable request.

Acknowledgments: MF was supported by the Fondazione Umberto Veronesi. The authors wish to thank Marco De Vincentiis for language editing.

Conflicts of Interest: The authors declare no conflict of interest.

References

1. De Lorenzo, A.; Romano, L.; Di Renzo, L.; Di Lorenzo, N.; Cenname, G.; Gualtieri, P. Obesity: A preventable, treatable, but relapsing disease. *Nutrition* **2020**, *71*, 110615. [CrossRef] [PubMed]
2. Blüher, M.; Aras, M.; Aronne, L.J.; Batterham, R.; Giorgino, F.; Ji, L.; Pietiläinen, K.H.; Schnell, O.; Tonchevska, E.; Wilding, J.P.H. New insights into the treatment of obesity. *Diabetes Obes. Metab.* **2023**, *25*, 2058–2072. [CrossRef] [PubMed]
3. Kiani, A.K.; Bonetti, G.; Donato, K.; Kaftalli, J.; Herbst, K.L.; Stuppia, L.; Fioretti, F.; Nodari, S.; Perrone, M.; Chiurazzi, P.; et al. Polymorphisms, diet and nutrigenomics. *J. Prev. Med. Hyg.* **2022**, *63* (Suppl. S3), E125–E141.
4. Di Renzo, L.; Gualtieri, P.; Romano, L.; Marrone, G.; Noce, A.; Pujia, A.; Perrone, M.A.; Aiello, V.; Colica, C.; De Lorenzo, A. Role of Personalized Nutrition in Chronic-Degenerative Diseases. *Nutrients* **2019**, *11*, 1707. [CrossRef] [PubMed]

5. Valenzuela, F.J.; Vera, J.; Venegas, C.; Muñoz, S.; Oyarce, S.; Muñoz, K.; Lagunas, C. Evidences of Polymorphism Associated with Circadian System and Risk of Pathologies: A Review of the Literature. *Int. J. Endocrinol.* **2016**, *2016*, 2746909. [CrossRef] [PubMed]
6. Corella, D.; Asensio, E.M.; Coltell, O.; Sorlí, J.V.; Estruch, R.; Martínez-González, M.Á. CLOCK gene variation is associated with incidence of type-2 diabetes and cardiovascular diseases in type-2 diabetic subjects: Dietary modulation in the PREDIMED randomized trial. *Cardiovasc. Diabetol.* **2016**, *15*, 4. [CrossRef] [PubMed]
7. Jagannath, A.; Taylor, L.; Wakaf, Z.; Vasudevan, S.R.; Foster, R.G. The genetics of circadian rhythms, sleep and health. *Hum. Mol. Genet.* **2017**, *26*, R128–R138. [CrossRef]
8. Angelousi, A.; Kassi, E.; Nasiri-Ansari, N.; Weickert, M.O.; Randeva, H.; Kaltsas, G. Clock genes alterations and endocrine disorders. *Eur. J. Clin. Investig.* **2018**, *48*, e12927. [CrossRef]
9. Scott, E.M.; Carter, A.M.; Grant, P.J. Association between polymorphisms in the Clock gene, obesity and the metabolic syndrome in man. *Int. J. Obes.* **2007**, *32*, 658–662. [CrossRef]
10. Franzago, M.; Alessandrelli, E.; Notarangelo, S.; Stuppia, L.; Vitacolonna, E. Chrono-Nutrition: Circadian Rhythm and Personalized Nutrition. *Int. J. Mol. Sci.* **2023**, *24*, 2571. [CrossRef]
11. Garaulet, M.; Sanchez-Moreno, C.; Smith, C.E.; Lee, Y.C.; Nicolas, F.; Ordovas, J.M. Ghrelin, sleep reduction and evening preference: Relationships to CLOCK 3111 T/C SNP and weight loss. *PLoS ONE* **2011**, *6*, e17435. [CrossRef] [PubMed]
12. Jaime-Lara, R.B.; Brooks, B.E.; Vizioli, C.; Chiles, M.; Nawal, N.; Ortiz-Figueroa, R.S.E.; Livinski, A.A.; Agarwal, K.; Colina-Prisco, C.; Iannarino, N.; et al. A systematic review of the biological mediators of fat taste and smell. *Physiol. Rev.* **2023**, *103*, 855–918. [CrossRef] [PubMed]
13. Abumrad, N.A. CD36 may determine our desire for dietary fats. *J. Clin. Investig.* **2005**, *115*, 2965–2967. [CrossRef] [PubMed]
14. Keller, K.L.; Liang, L.C.; Sakimura, J.; May, D.; van Belle, C.; Breen, C.; Driggin, E.; Tepper, B.J.; Lanzano, P.C.; Deng, L.; et al. Common variants in the CD36 gene are associated with oral fat perception, fat preferences, and obesity in African Americans. *Obes. Silver Spring* **2012**, *20*, 1066–1073. [CrossRef]
15. Pepino, M.Y.; Love-Gregory, L.; Klein, S.; Abumrad, N.A. The fatty acid translocase gene CD36 and lingual lipase influence oral sensitivity to fat in obese subjects. *J. Lipid Res.* **2012**, *53*, 561–566. [CrossRef]
16. Reed, D.R.; Xia, M.B. Recent advances in fatty acid perception and genetics. *Adv. Nutr.* **2015**, *6*, 353s–360s. [CrossRef]
17. Karmous, I.; Plesnik, J.; Khan, A.S.; Sery, O.; Abid, A.; Mankai, A.; Aouidet, A.; Khan, N.A. Orosensory detection of bitter in fat-taster healthy and obese participants: Genetic polymorphism of CD36 and TAS2R38. *Clin. Nutr.* **2017**, *37*, 313–320. [CrossRef]
18. Melis, M.; Sollai, G.; Muroni, P.; Crnjar, R.; Barbarossa, I.T. Associations between orosensory perception of oleic acid, the common single nucleotide polymorphisms (rs1761667 and rs1527483) in the CD36 gene, and 6-n-propylthiouracil (PROP) tasting. *Nutrients* **2015**, *7*, 2068–2084. [CrossRef]
19. Mrizak, I.; Sery, O.; Plesnik, J.; Arfa, A.; Fekih, M.; Bouslema, A.; Zaouali, M.; Tabka, Z.; Khan, N.A. The A allele of cluster of differentiation 36 (CD36) SNP 1761667 associates with decreased lipid taste perception in obese Tunisian women. *Br. J. Nutr.* **2015**, *113*, 1330–1337. [CrossRef]
20. Burgess, B.; Melis, M.; Scoular, K.; Driver, M.; Schaich, K.M.; Keller, K.L.; Tomassini Barbarossa, I.; Tepper, B.J. Effects of CD36 Genotype on Oral Perception of Oleic Acid Supplemented Safflower Oil Emulsions in Two Ethnic Groups: A Preliminary Study. *J. Food Sci.* **2018**, *83*, 1373–1380. [CrossRef]
21. Love-Gregory, L.; Sherva, R.; Schappe, T.; Qi, J.S.; McCrea, J.; Klein, S.; Connelly, M.A.; Abumrad, N.A. Common CD36 SNPs reduce protein expression and may contribute to a protective atherogenic profile. *Hum. Mol. Genet.* **2011**, *20*, 193–201. [CrossRef] [PubMed]
22. Ghosh, A.; Murugesan, G.; Chen, K.; Zhang, L.; Wang, Q.; Febbraio, M.; Anselmo, R.M.; Marchant, K.; Barnard, J.; Silverstein, R.L. Platelet CD36 surface expression levels affect functional responses to oxidized LDL and are associated with inheritance of specific genetic polymorphisms. *Blood* **2011**, *117*, 6355–6366. [CrossRef] [PubMed]
23. Melis, M.; Carta, G.; Pintus, S.; Pintus, P.; Piras, C.A.; Murru, E.; Manca, C.; Di Marzo, V.; Banni, S. Tomassini Barbarossa, I. Polymorphism rs1761667 in the CD36 Gene Is Associated to Changes in Fatty Acid Metabolism and Circulating Endocannabinoid Levels Distinctively in Normal Weight and Obese Subjects. *Front. Physiol.* **2017**, *8*, 1006. [CrossRef] [PubMed]
24. Yazdanpanah, Z.; Salehi-Abargouei, A.; Mollahosseini, M.; Sheikhha, M.H.; Mirzaei, M.; Mozaffari-Khosravi, H. The cluster of differentiation 36 (*CD36*) rs1761667 polymorphism interacts with dietary patterns to affect cardiometabolic risk factors and metabolic syndrome risk in apparently healthy individuals. *Br. J. Nutr.* **2023**, *130*, 1510–1520. [CrossRef] [PubMed]
25. Trento, M.; Passera, P.; Borgo, E.; Tomalino, M.; Bajardi, M.; Cavallo, F.; Porta, M. A 5-year randomized controlled study of learning, problem solving ability, and quality of life modifications in people with type 2 diabetes managed by group care. *Diabetes Care* **2004**, *27*, 670–675. [CrossRef]
26. Wang, Y.; Xue, H.; Huang, Y.; Huang, L.; Zhang, D.A. Systematic review of application and effectiveness of health interventions for obesity and diabetes treatment and self-management. *Adv. Nutr.* **2017**, *8*, 449–462. [CrossRef]
27. Trento, M.; Passera, P.; Tomalino, M.; Bajardi, M.; Pomero, F.; Allione, A.; Vaccari, P.; Psych, B. Group visits improve metabolic control in type 2 diabetes: A 2-year follow-up. *Diabetes Care* **2001**, *24*, 995–1000. [CrossRef]
28. Trento, M.; Passera, P.; Bajardi, M.; Tomalino, M.; Grassi, G.; Borgo, E.; Donnola, C.; Cavallo, F.; Bondonio, P.; Porta, M. Lifestyle intervention by group care prevents deterioration of type ii diabetes: A 4-year randomized controlled clinical trial. *Diabetologia* **2002**, *45*, 1231–1239. [CrossRef]

29. Franzago, M.; Di Nicola, M.; Fraticelli, F.; Marchioni, M.; Stuppia, L.; Vitacolonna, E. Nutrigenetic variants and response to diet/lifestyle intervention in obese subjects: A pilot study. *Acta Diabetol.* **2022**, *59*, 69–81. [CrossRef]
30. Evert, A.B.; Dennison, M.; Gardner, C.D.; Garvey, W.T.; Lau, K.H.K.; MacLeod, J.; Mitri, J.; Pereira, R.F.; Rawlings, K.; Saslow, L.; et al. Nutrition Therapy for Adults with Diabetes or Prediabetes: A Consensus Report. *Diabetes Care* **2019**, *42*, 731–754. [CrossRef]
31. Natale, V.; Esposito, M.J.; Martoni, M.; Fabbri, M. Validity of the reduced version of the Morningness-Eveningness Questionnaire. *Sleep Biol. Rhythm.* **2006**, *4*, 72–74. [CrossRef]
32. Martínez-González, M.A.; García-Arellano, A.; Toledo, E.; Salas-Salvadó, J.; Buil-Cosiales, P.; Corella, D.; Covas, M.I.; Schröder, H.; Arós, F.; Gómez-Gracia, E.; et al. A 14-item Mediterranean diet assessment tool and obesity indexes among high risk subjects: The PREDIMED trial. *PLoS ONE* **2012**, *7*, e43134. [CrossRef] [PubMed]
33. Mannocci, A.; Di Thiene, D.; Del Cimmuto, A.; Masala, D. International physical activity questionnaire: Validation and assessment in an Italian sample. *Ital. J. Public Health* **2010**, *7*, 369–376.
34. Gautam, S.; Agrawal, C.G.; Banerjee, M. CD36 gene variants in early prediction of type 2 diabetes mellitus. *Genet. Test. Mol. Biomark.* **2015**, *19*, 144–149. [CrossRef]
35. Sayed, A.; Šerý, O.; Plesnik, J.; Daoudi, H.; Rouabah, A.; Rouabah, L.; Khan, N.A. CD36 AA genotype is associated with decreased lipid taste perception in young obese, but not lean, children. *Int. J. Obes.* **2015**, *39*, 920–924. [CrossRef]
36. Sookoian, S.; Gemma, C.; Gianotti, T.F.; Burgueño, A.; Castaño, G.; Pirola, C.J. Genetic variants of Clock transcription factor are associated with individual susceptibility to obesity. *Am. J. Clin. Nutr.* **2008**, *87*, 1606–1615. [CrossRef]
37. Lopez-Minguez, J.; Gómez-Abellán, P.; Garaulet, M. Circadian rhythms, food timing and obesity. *Proc. Nutr. Soc.* **2016**, *75*, 501–511. [CrossRef]
38. Rahati, S.; Qorbani, M.; Naghavi, A.; Nia, M.H.; Pishva, H. Association between CLOCK 3111 T/C polymorphism with ghrelin, GLP-1, food timing, sleep and chronotype in overweight and obese Iranian adults. *BMC Endocr. Disord.* **2022**, *22*, 147. [CrossRef]
39. Woon, P.Y.; Kaisaki, P.J.; Bragança, J.; Bihoreau, M.-T.; Levy, J.C.; Farrall, M.; Gauguier, D. Aryl hydrocarbon receptor nuclear translocator-like (BMAL1) is associated with susceptibility to hypertension and type 2 diabetes. *Proc. Natl. Acad. Sci. USA* **2007**, *104*, 14412–14417. [CrossRef]
40. Uemura, H.; Katsuura-Kamano, S.; Yamaguchi, M.; Arisawa, K.; Hamajima, N.; Hishida, A.; Kawai, S.; Oze, I.; Shinchi, K.; Takashima, N.; et al. Variant of the clock circadian regulator (CLOCK) gene and related haplotypes are associated with the prevalence of type 2 diabetes in the Japanese population. *J. Diabetes* **2015**, *8*, 667–676. [CrossRef]
41. Stevens, H.; Verdone, G.; Lang, L.; Graham, C.; Pilic, L.; Mavrommatis, Y. A systematic review of variations in circadian rhythm genes and type 2 diabetes. *Nutr. Health.* **2023**. [CrossRef] [PubMed]
42. Garaulet, M.; Lee, Y.-C.; Shen, J.; Parnell, L.D.; Arnett, D.K.; Tsai, M.Y.; Lai, C.-Q.; Ordovas, J.M. CLOCK genetic variation and metabolic syndrome risk: Modulation by monounsaturated fatty acids. *Am. J. Clin. Nutr.* **2009**, *90*, 1466–1475. [CrossRef] [PubMed]
43. Garaulet, M.; Corbalán, M.D.; Madrid, J.A.; Morales, E.; Baraza, J.C.; Lee, Y.C.; Ordovas, J.M. CLOCK gene is implicated in weight reduction in obese patients participating in a dietary programme based on the Mediterranean diet. *Int. J. Obes.* **2010**, *34*, 516–523. [CrossRef] [PubMed]
44. Zhao, L.; Li, Y.; Ding, Q.; Li, Y.; Chen, Y.; Ruan, X.Z. CD36 senses dietary lipids and regulates lipids homeostasis in the intestine. *Front. Physiol.* **2021**, *12*, 516–523. [CrossRef]
45. Abdoul-Azize, S.; Selvakumar, S.; Sadou, H.; Besnard, P.; Khan, N.A. Ca2þ signaling in taste bud cells and spontaneous preference for fat: Unresolved roles of CD36 and GPR120. *Biochimie* **2014**, *96*, 8e13. [CrossRef]
46. Zhang, X.J.; Zhou, L.H.; Ban, X.; Liu, D.X.; Jiang, W.; Liu, X.M. Decreased expression of CD36 in circumvallate taste buds of high-fat diet induced obese rats. *Acta Histochem.* **2011**, *113*, 663–667. [CrossRef]
47. Karthi, M.; Deepankumar, S.; Vinithra, P.; Gowtham, S.; Vasanth, K.; Praveen Raj, P.; Senthilkumar, R.; Selvakumar, S. Single nucleotide polymorphism in CD36: Correlation to peptide YY levels in obese and non-obese adults. *Clin. Nutr.* **2021**, *40*, 2707–2715. [CrossRef]
48. Zhang, Y.; Ling, Z.Y.; Deng, S.B.; Du, H.A.; Yin, Y.H.; Yuan, J.; She, Q.; Chen, Y.Q. Associations between CD36 gene polymorphisms and susceptibility to coronary artery heart disease. *Braz. J. Med. Biol. Res.* **2014**, *47*, 895–903. [CrossRef]
49. Boghdady, A.; Arafa, U.A.; Sabet, E.A.; Salama, E.; El Sharawy, A.; Elbadry, M.I. Association between rs1761667 polymorphism of CD36 gene and risk of coronary atherosclerosis in Egyptian population. *Cardiovasc. Diagn. Ther.* **2016**, *6*, 120–130. [CrossRef]
50. Banerjee, M.; Gautam, S.; Saxena, M.; Bid, H.K.; Agrawal, C.G. Association of CD36 gene variants rs1761667 (G > A) and rs1527483 (C > T) with Type 2 diabetes in North Indian population. *Int. J. Diabetes Mellit.* **2010**, *2*, 179–183. [CrossRef]
51. Bayoumy, N.M.; El-Shabrawi, M.M.; Hassan, H.H. Association of cluster of differentiation 36 gene variant rs1761667 (G > A) with metabolic syndrome in Egyptian adults. *Saudi Med. J.* **2012**, *33*, 489–494. [PubMed]
52. Pioltine, M.B.; de Melo, M.E.; Santos, A.; Machado, A.D.; Fernandes, A.E.; Fujiwara, C.T.; Cercato, C.; Mancini, M.C. Genetic variation in CD36 is associated with decreased fat and sugar intake in obese children and adolescents. *J Nutr. Nutr.* **2016**, *9*, 300–305. [CrossRef] [PubMed]
53. Solakivi, T.; Kunnas, T.; Nikkari, S.T. Contribution of fatty acid transporter (CD36) genetic variant rs1761667 to body mass index, the TAMRISK study. *Scand. J. Clin. Lab. Investig.* **2015**, *75*, 254–258. [CrossRef] [PubMed]
54. Luangrath, V.; Brodeur, M.R.; Rhainds, D.; Brissette, L. Mouse CD36 has opposite effects on LDL and oxidized LDL metabolism in vivo. *Arterioscler. Thromb. Vasc. Biol.* **2008**, *28*, 1290–1295. [CrossRef] [PubMed]

55. Brundert, M.; Heeren, J.; Merkel, M.; Carambia, A.; Herkel, J.; Groitl, P.; Dobner, T.; Ramakrishnan, R.; Moore, K.J.; Rinninger, F. Scavenger receptor CD36 mediates uptake of high density lipoproteins in mice and by cultured cells. *J. Lipid Res.* **2011**, *52*, 745–758. [CrossRef]
56. Noel, S.E.; Lai, C.Q.; Mattei, J.; Parnell, L.D.; Ordovas, J.M.; Tucker, K.L. Variants of the CD36 gene and metabolic syndrome in Boston Puerto Rican adults. *Atherosclerosis* **2010**, *211*, 210–215. [CrossRef]
57. Yazdanpanah, Z.; Mozaffari-Khosravi, H.; Mirzaei, M.; Sheikhha, M.H.; Salehi-Abargouei, A. A systematic review and meta-analysis on the association between CD36 rs1761667 polymorphism and cardiometabolic risk factors in adults. *Sci. Rep.* **2022**, *12*, 5916. [CrossRef]
58. Rać, M.; Krupa, B.; Garanty-Bogacka, B.; Syrenicz, M.; Safranow, K.; Dziedziejko, V.; Kurzawski, G.; Olszewska, M.; Rać, M.; Chlubek, D. Polymorphism of CD36 gene, carbohydrate metabolism and plasma CD36 concentration in obese children. A preliminary study. *Postep. Hig. Med. Dosw. Online* **2012**, *66*, 954–958. [CrossRef]
59. Julious, S.A. Sample size of 12 per group rule of thumb for a pilot study. *Pharm. Stat.* **2005**, *4*, 287–291. [CrossRef]
60. Moore, C.; Nietert, S. Recommendations for planning pilot studies in clinical and translational sciences. *Clin. Transl. Sci.* **2011**, *4*, 332–337. [CrossRef]

nutrients MDPI

Article

Tributyrin Mitigates Ethanol-Induced Lysine Acetylation of Histone-H3 and p65-NFκB Downregulating CCL2 Expression and Consequent Liver Inflammation and Injury

Smita S. Ghare [1,2,3,*], Benjamin T. Charpentier [2,4], Dushan T. Ghooray [1,2,3], Jingwen Zhang [1,2,3], Manicka V. Vadhanam [1,2,3], Sreelatha Reddy [1,2,3], Swati Joshi-Barve [1,2,3,†], Craig J. McClain [1,2,3,5] and Shirish S. Barve [1,2,3]

1 Department of Medicine, University of Louisville, Louisville, KY 40202, USA
2 UofL Alcohol Center, University of Louisville, Louisville, KY 40202, USA
3 UofL Hepatobiology COBRE, University of Louisville, Louisville, KY 40202, USA
4 Department of Anatomical Science and Neurobiology, University of Louisville, Louisville, KY 40202, USA
5 Robley Rex VA Medical Center, University of Louisville, Louisville, KY 40202, USA
* Correspondence: smita.ghare@louisville.edu; Tel.: +1-(502)-852-9221; Fax: +1-(502)-852-8927
† Passed away.

Abstract: Purpose: Chemokine-driven leukocyte infiltration and sustained inflammation contribute to alcohol-associated liver disease (ALD). Elevated hepatic CCL2 expression, seen in ALD, is associated with disease severity. However, mechanisms of CCL2 regulation are not completely elucidated. Post-translational modifications (PTMs) of proteins, particularly acetylation, modulate gene expression. This study examined the acetylation changes of promoter-associated histone-H3 and key transcription factor-NFκB in regulating hepatic CCL2 expression and subsequent inflammation and injury. Further, the effect of therapeutic modulation of the acetylation state by tributyrin (TB), a butyrate prodrug, was assessed. Methods: Hepatic CCL2 expression was assessed in mice fed control (PF) or an ethanol-containing Lieber–DeCarli (5% v/v, EF) diet for 7 weeks with or without oral administration of tributyrin (TB, 2 g/kg, 5 days/week). A chromatin immunoprecipitation (ChIP) assay evaluated promoter-associated modifications. Nuclear association between SIRT1, p300, and NFκB-p65 and acetylation changes of p65 were determined using immunoprecipitation and Western blot analyses. A Student's t-test and one-way ANOVA determined the significance. Results: Ethanol significantly increased promoter-associated histone-H3-lysine-9 acetylation (H3K9Ac), reflecting a transcriptionally permissive state with a resultant increase in hepatic CCL2 mRNA and protein expression. Moreover, increased lysine-310-acetylation of nuclear RelA/p65 decreased its association with SIRT1, a class III HDAC, but concomitantly increased with p300, a histone acetyltransferase. This further led to enhanced recruitment of NF-κB/p65 and RNA polymerase-II to the CCL2 promoter. Oral TB administration prevented ethanol-associated acetylation changes, thus downregulating CCL2 expression, hepatic neutrophil infiltration, and inflammation/ injury. Conclusion: The modulation of a protein acetylation state via ethanol or TB mechanistically regulates hepatic CCL2 upregulation in ALD.

Keywords: alcohol; histone H3 acetylation; p300; SIRT1; CCL2 promoter; tributyrin

1. Introduction

Sustained hepatic inflammation is an important factor in alcohol-associated liver disease (ALD) development and progression. Neutrophil infiltration and monocyte recruitment in the liver are associated with chronic alcohol-induced liver inflammation and injury [1,2]. Chemokines act as chemo-attractants and activators, allowing the recruitment of these neutrophils and monocytes to the liver and inducing an innate and adaptive immune response [3,4]. Among four different chemokine families, the C-C subfamily is the largest, and chemokines such as C-C motif chemokine ligand 2 (CCL2), also referred to as monocyte chemotactic peptide-1 (MCP-1), are reported to be present in high levels in patients with alcohol-associated hepatitis [5,6]. Both plasma levels and hepatic expression of CCL2 have been reported to correlate with disease severity [5]. Importantly, CCL2 knock-out mice were protected against alcoholic liver injury by the induction of genes responsible for fatty acid oxidation and inhibition of pro-inflammatory cytokines [7]. Although these studies emphasize the key role of CCL2 in inducing liver pathologies, including hepatic inflammation, the molecular mechanisms that regulate their expression in ALD, particularly in response to alcohol, are yet to be completely elucidated.

Emerging studies have documented the role of the gut–liver axis in ALD in the context of alcohol-induced alterations in gut microbiota, barrier dysfunction, and endotoxemia leading to sustained hepatic inflammation. Alcohol consumption is known to affect short-chain fatty acids (SCFAs) levels, particularly butyrate [8]. Butyrate has been demonstrated to have an epigenetic regulatory role by acting as an HDAC inhibitor [9–11]. Additionally, butyrate has been shown to inhibit inflammatory responses [12–14]. Tributyrin is a butyrate prodrug consisting of three butyrate molecules that can be released in the circulation upon digestion by gastric and pancreatic amylases and lipases. Experimental studies have shown that tributyrin supplementation exhibits beneficial effects via maintaining gut bacterial diversity and barrier integrity in ethanol-fed animals [15]. Moreover, our previous work showed that oral administration of tributyrin can protect against ethanol-induced alterations in hepatic HDACs expression/activity and promoter histone modifications and can regulate the expression of genes involved in lipid metabolism [16]. Since there is no specific FDA-approved treatment for ALD and novel therapies are urgently needed, the potential role of tributyrin administration as a dietary therapeutic option against sustained hepatic inflammation needs to be investigated.

Epigenetic regulation is critical for gene expression. In particular, a post-translational acetylation modification is known to dictate the transcriptional activation of a given gene, as this modification regulates functions of both histones and non-histone proteins, like transcription factors [17]. Histone deacetylases (HDACs) and histone acetyltransferases (HATs) maintain the acetylation states of these proteins and play a key role in gene transcription [18,19]. We have shown that alcohol alters hepatic HDAC expression in a binge alcohol model [20]. Alcohol is also known to increase the global histone acetylation levels in the liver [21–24]; however, limited information is available for gene-specific effects.

In the present study, we demonstrated that ethanol-induced pathogenic alterations in acetylation states of promoter-associated histone H3 and essential transcription factor NFκB contribute to the upregulation of hepatic CCL2 expression. Further, tributyrin, a dietary butyrate prodrug, prevents ethanol-mediated modulation of SIRT-1-NFκB-p300 interactions and resultant acetylation changes.

2. Materials and Methods

Animal Model: Eight-week-old male C57BL/6N mice were obtained from Harlan (Indianapolis, IN, USA). All mice were housed in a pathogen-free, temperature-controlled animal facility accredited by the Association for Assessment and Accreditation of Laboratory Animal Care with 12 h light/12 h dark cycles. All experiments were carried out according to the criteria outlined in the Guide for Care and Use of Laboratory Animals and with the approval from the University of Louisville Animal Care and Use Committee (IACUC# 20754). Mice were fed a modified Lieber–DeCarli liquid diet enriched in unsaturated fat (corn oil), which provided 40% of energy from fat, 43% from carbohydrate, and 17% from protein (Bio-Serv Inc., Frenchtown, NJ, USA). Mice were pair-fed a Lieber–DeCarli liquid diet containing either ethanol (EF) or isocaloric maltose dextrin (PF) for 7 weeks. Ethanol was gradually increased for one week, and then, mice were fed the ethanol diet (5% (v/v)) ad libitum for 7 weeks (EF). The control-pair-fed (PF) mice were given the isocaloric maltose-dextrin-containing liquid diet. For tributyrin treatment groups, both pair-fed (TB) and ethanol-fed (EF + TB) mice received tributyrin (Sigma Aldrich, St.

Louis, MO, USA) by oral gavage (2 g/kg, 5 days per week) for 7 weeks as a preventive strategy (EF + TB (7 + 7)) or for the last 3 weeks as an interventional strategy (EF + TB (7 + 3)). The tributyrin-alone group (TB) received oral TB administration (2 g/kg, 5 days per week) for 7 weeks.

Liver Histopathological Examination: For histological analysis, liver sections were fixed in 10% buffered formalin for 24 h and embedded in paraffin. Tissue sections were stained with hematoxylin-eosin (H&E), F4/80+, and CAE staining and examined under light microscopy with 20× magnification. Assays were performed according to the manufacturer's protocols.

Assessment of Liver Injury: The liver injury was assessed by measuring alanine aminotransferase (ALT) and aspartate aminotransferase (AST) using commercially available kits and manufacturer's instructions (Thermo Electron, Melbourne, Australia).

Blood Endotoxin Assay: Serum endotoxin levels were measured using a Limulus Amebocyte Lysate kit (Lonza, Walkerville, MD, USA) according to the manufacturer's instructions.

MPO Assay: For hepatic myeloperoxidase activity (MPO), liver samples were homogenized (10 μg/200 μL) in 200 mM NaCl, 5 mM EDTA, 10 mM tris, and 10% glycerin and centrifuged at 1500× *g* for 15 min. The MPO levels were measured according to the manufacturer's instructions (Hycult Biotech, Plymouth Meeting, PA, USA) at 450 nm using a plate reader. Results are expressed as mg of MPO levels/gram of protein, as determined via a Lowry protein assay (Bio-Rad, Hercules, CA, USA)

Hepatic CCL2 Protein Measurements: Liver lysates were homogenized in a lysis buffer (20 mM Tris, 0.25 M sucrose, 2 mM EDTA, 10 mM EGTA, and 1% Triton X-100), and analysis was performed using a U-Plex kit from Mesoscale Discovery (MSD) (Rockville, ML, USA). Over 200 μg of protein was loaded into each well. The plate was read using a MESO QuickPlex SQ 120 imager and analyzed using Discovery Workbench v4.0 software. The assay was performed according to the manufacturer's instructions.

RNA Isolation and Real-Time PCR Analysis: Total RNA was isolated using TRIzol (Invitrogen, Carlsbad, CA, USA), and cDNA was made using Quanta qScript (Quanta BioSciences, Gaithersburg, MD, USA). The real-time PCR was performed with Quanta Perfecta SYBR green fast mix and an ABI Prism 7500 sequence detection system (Applied Biosystems, Foster City, CA, USA). The relative gene expression was analyzed using the $2^{-\Delta\Delta Ct}$ method by normalizing with TBP (TATA-binding protein) gene expression in all the experiments and is presented as a fold change over untreated/pair-fed, which was set at 1.

Mouse CCL2 mRNA primers

Forward: GGCTCAGCCAGATGCAGT

Reverse: TGAGCTTGGTGACAAAAACTACAG

Mouse TBP mRNA primers

Forward: CCTAAAGACCATTGCACTTCGT

Reverse: GCTCCTGTGCACACCATTTT

Chromatin Immunoprecipitation (ChIP) and qChIP PCR Analysis: The ChIP assay was conducted using the ChIP assay protocol established in the lab [25]. ChIP antibodies directed against anti-acetyl H3K9 (17-615), anti-phospho H3S10 (17-685), anti-p300 (05-257), anti-p65 (17-10060), and anti-RNA polymerase II (17-672) were purchased from EMD Millipore (Burlington, MA, USA). ChIP-PCR primers designed for the regions of the CCL2 were used, and their sequences are detailed below. Data were analyzed as a differential occupancy fold change. ChIP-qPCR results were calculated using the ΔΔCt method, where each ChIP DNA fraction's Ct value was normalized to the input DNA fraction. The specificity of each ChIP was established using the corresponding isotype-specific control antibodies (IgG).

Three ChIP-PCR primers were designed for each region of the mouse-CCL2 promoter. The primer sequences are as follows:

(Region I) mmCcl2_F1	CAAGCACCCTGCCTGACT
(Region I) mmCcl2_R1	CTCCCGTCTGGCTCTCTG
(Region II) mmCcl2_F2	TCCCAGGAGTGGCTAGAAAA
(Region II) mmCcl2_R2	TCCGCTGAGTAAGTGCAGAG
(Region III) mmCcl2_F3	CATCTGGAGCTCACATTCCA
(Region III) mmCcl2_R3	GGCAGGTCAGAGGCAGAGTA

Immunoprecipitation (IP) and Western Blot (WB) Analysis: Total nuclear extracts were prepared by lysing liver tissue in a RIPA lysis buffer (25 mM Tris/HCl pH 7.5150 mM NaCl, 1% NP-40, 0.1% SDS, 1 mM DTT, 10% glycerol, 1× protease inhibitor cocktail, 0.5% sodium deoxycholate, 1 mM Na_2VO_3, and 10 mM NaF). All IP and WB analyses were performed as described earlier [26]. Detection of Lamine-B1 served as a loading control. Quantification was performed with ImageLab analysis software v6.0.1 (Bio-Rad). The data shown are representative of three separate experiments showing similar results. Primary antibodies for Acetyl-NF-κB p65 (Lys310) (#3045), SIRT1 (#2310), p300 (#86377), NF-κB p65 (L8F6, mouse mAb #6956), NF-κB p65 (D14E12, rabbit mAb #8242), GAPDH (D16H11, #5174), Lamin B1 (D9V6H, #13435), and secondary antibodies anti-rabbit IgG, HRP-linked antibody (#7074), and anti-mouse IgG, HRP-linked antibody (#7076) were purchased from Cell Signaling Technology, Inc. (Danvers, MA, USA).

Statistical Analysis: Data are presented as mean ± SEM for the indicated number of independently performed experiments or 4–6 mice per group. A Student's *t*-test and a one-way ANOVA with Bonferroni multiple comparison tests were used to determine statistical significance. A $p < 0.05$ was considered statistically significant.

All authors had access to all data and reviewed and approved the final manuscript.

3. Results

3.1. Oral Administration of Tributyrin Attenuates Alcohol-Induced, Neutrophil Infiltration, Kupffer Cell Activation, Systemic Endotoxemia, and Hepatic Injury

Both preventive and interventional treatment strategies were employed to test the effect of oral administration of tributyrin on alcohol-induced hepatic neutrophil infiltration and Kupffer cell activation in experimental ALD. Initially, a histological assessment was performed to examine the effect of tributyrin administration on markers of hepatic inflammation and injury. In accordance with earlier reports, livers from mice fed ethanol for 7 weeks (EF) had significantly higher hepatic steatosis, as shown by H&E staining (Figure 1A) and neutrophil infiltration, as indicated by CAE staining (Figure 1B). Importantly, tributyrin administration, by both preventive and interventional strategies, inhibited ethanol-induced hepatic steatosis (Figure 1A) and neutrophil infiltration of liver tissue (Figure 1B). Moreover, there was a significant decrease in the hepatic levels of myeloperoxidase (MPO), an enzyme expressed predominantly in neutrophils (Figure 1C). Additionally, as compared to livers from ethanol-fed mice, there was a significant reduction in the number of activated macrophages as seen by markedly fewer F4/80 positive (brown color-stained cells) Kupffer cells in mice administered TB (Figure 1D) with the concomitant decrease in systemic endotoxin levels (Figure 1E) and markers of liver injury (AST/ALT ratio) (Figure 1F).

3.2. Tributyrin Mitigates Ethanol-Inducible Increase in Hepatic CCL2 Chemokine Expression

Chemokines act as a critical mediator for the recruitment of neutrophils and macrophages, and elevated expression of hepatic CCL2 has been well documented to correlate with PMN infiltration and disease severity in ALD [5]. Hence, we next examined the effect of ethanol feeding and tributyrin administration on hepatic CCL2 expression. Our data showed that chronic ethanol feeding caused an increase in hepatic expression of CCL2 at both mRNA (Figure 2A) and protein levels (Figure 2B). Importantly, the increase in the hepatic CCL2 expression in response to ethanol was significantly prevented by tributyrin administration under both preventive and interventional strategies (Figure 2).

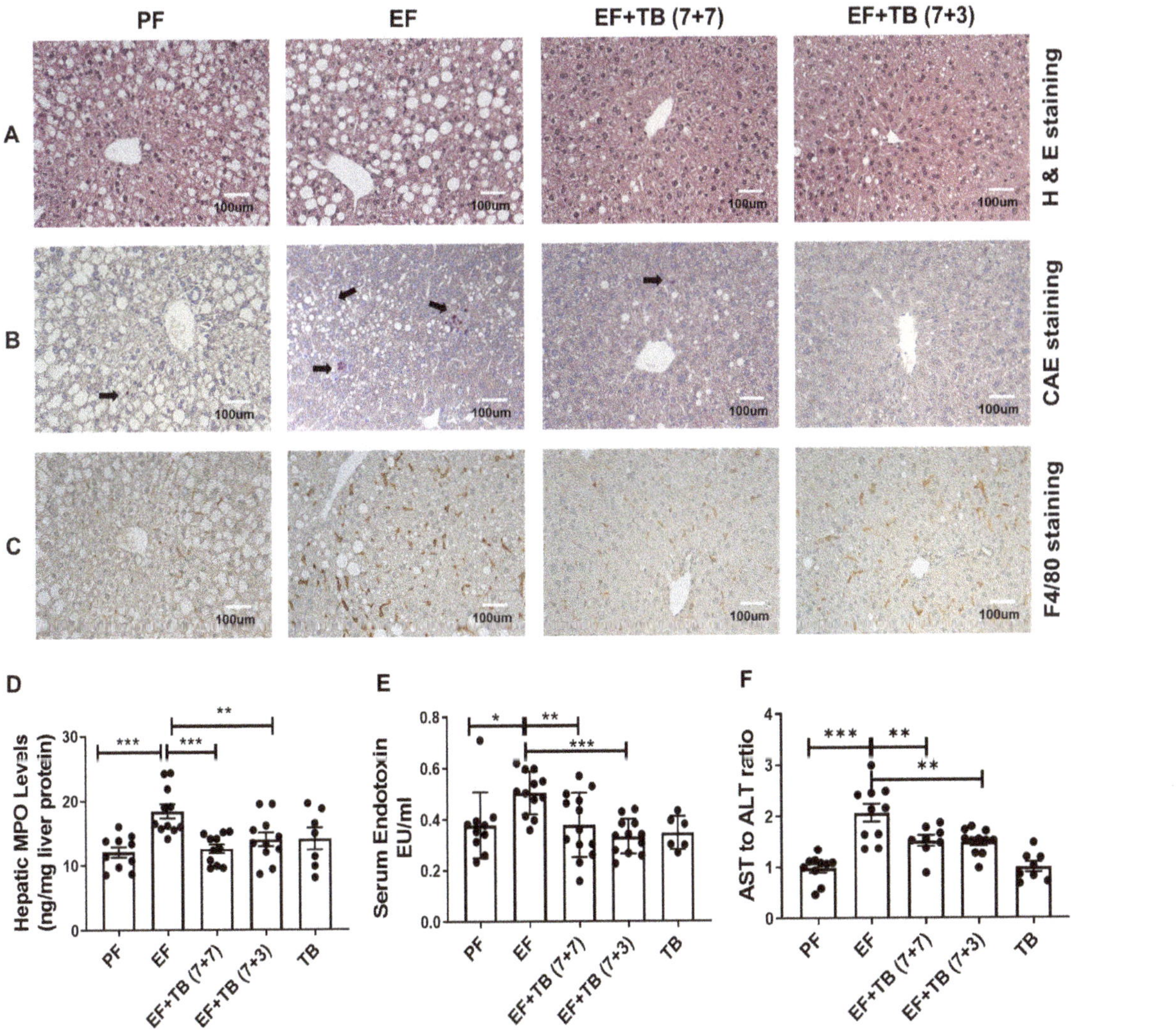

Figure 1. Effect of tributyrin administration on hepatic steatosis, inflammation, and injury in chronic ethanol-fed mice: Mice were fed either a control diet (Pair-fed, PF) or ethanol (ethanol-fed; EF) (5% *v*/*v*)-containing diet. Tributyrin (TB; 2 g/kg) was orally administered to ethanol-fed mice as a preventive (EF + TB (7 + 7)) or interventional (EF + TB (7 + 3)) treatment strategy. Immunohistochemical analysis was performed as (**A**) H&E for steatosis, (**B**) CAE staining for neutrophil infiltration and is indicated by black arrows, and (**C**) F4/80 staining for Kupffer cell activation in mice liver tissue sections frozen in OCT. All images were acquired using a 20× objective and scale bar showing 100 μm. (**D**) Hepatic MPO levels, (**E**) serum endotoxin levels, and the (**F**) AST to ALT ratio were biochemically assessed. Statistical analysis: mean ± SEM. * $p < 0.05$, ** $p < 0.01$, and *** $p < 0.001$ compared with PF or EF using ANOVA with Bonferroni's test (n = 8–10).

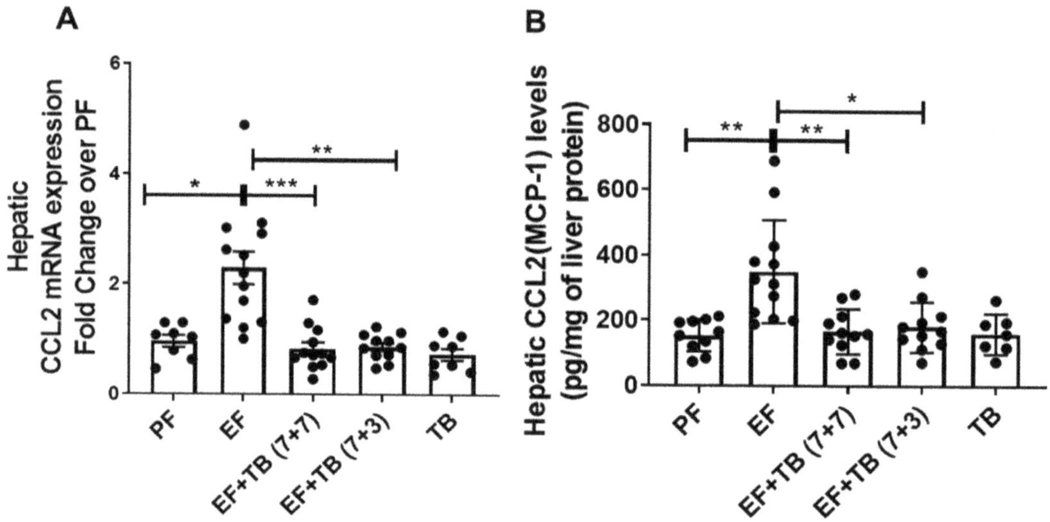

Figure 2. Effect of tributyrin administration on hepatic CCL2 chemokine gene expression in chronic ethanol-fed mice: Mice were fed either a control diet (pair-fed, PF) or ethanol (ethanol-fed; EF) (5% *v/v*)-containing diet. Tributyrin (TB; 2 g/kg) was orally administered to ethanol-fed mice as a preventive (EF + TB (7 + 7)) or interventional (EF + TB (7 + 3)) treatment strategy. (**A**) Hepatic mRNA analysis for CCL2 and TBP (TATA-binding protein) genes was performed using real-time PCR. CCL2 mRNA expression was normalized with TBP and presented as a fold change over PF. (**B**) Liver chemokine CCL2 (MCP-1) protein levels were analyzed using an MSD U-plex assay. The data are presented as a bar graph after normalizing CCL2 levels with total protein concentration. Statistical analysis: mean ± SEM. * $p < 0.05$, ** $p < 0.01$, and *** $p < 0.001$ compared with PF or EF using ANOVA with Bonferroni's test (n = 8–10).

3.3. Tributyrin Prevents Ethanol-Induced Transcriptionally Permissive CCL2 Promoter-Associated Histone Modifications in the Liver

Epigenetic modifications are known to orchestrate the interplay between the chromatin state and gene expression. To understand whether the post-translational histone modifications occurring at the CCL2 promoter contribute to its gene expression in response to alcohol and tributyrin, a Chromatin immunoprecipitation (ChIP) assay was performed. Promoter-associated histone modifications were investigated via ChIP qPCR at three transcriptionally relevant sites known for NFκB occupancy on the CCL2 promoter [27–29]. Region-I interrogated the TSS site (−66 to +19), and Region-II (−118 to −212) and Region-III (−2319 to −2431) investigated proximal and distal enhancer regions known for their NFκB binding on the CCL2 gene promoter (Figure 3A). The specificity of each ChIP was established using corresponding isotype-specific control antibodies (IgG)

The status of histone H3 lysine 9 acetylation (H3K9Ac), which plays a key role in the transcriptional activation of gene expression, was examined. ChIP analysis demonstrated that correspondent to CCL2 gene expression, ethanol feeding significantly increased transcriptionally permissive histone H3 lysine 9 acetylation (H3K9Ac) at all three regions of the CCL2 promoter in murine livers (Figure 3B). Additionally, the effect of ethanol was also examined on another transcriptionally permissive modification. Histone H3 serine 10 phosphorylation (H3S10P), which is also known to be linked with histone H3K9 acetylation was also examined (Figure 3C). The data showed that similar to H3K9 acetylation, ethanol feeding also increased H3S10 phosphorylation at all regions of the CCL2 promoter. In contrast, both preventive and interventional treatment strategies of tributyrin administration significantly inhibited ethanol-induced H3K9 acetylation and H3S10 phosphorylation at all regions of the CCL2 promoter (Figure 3B,C).

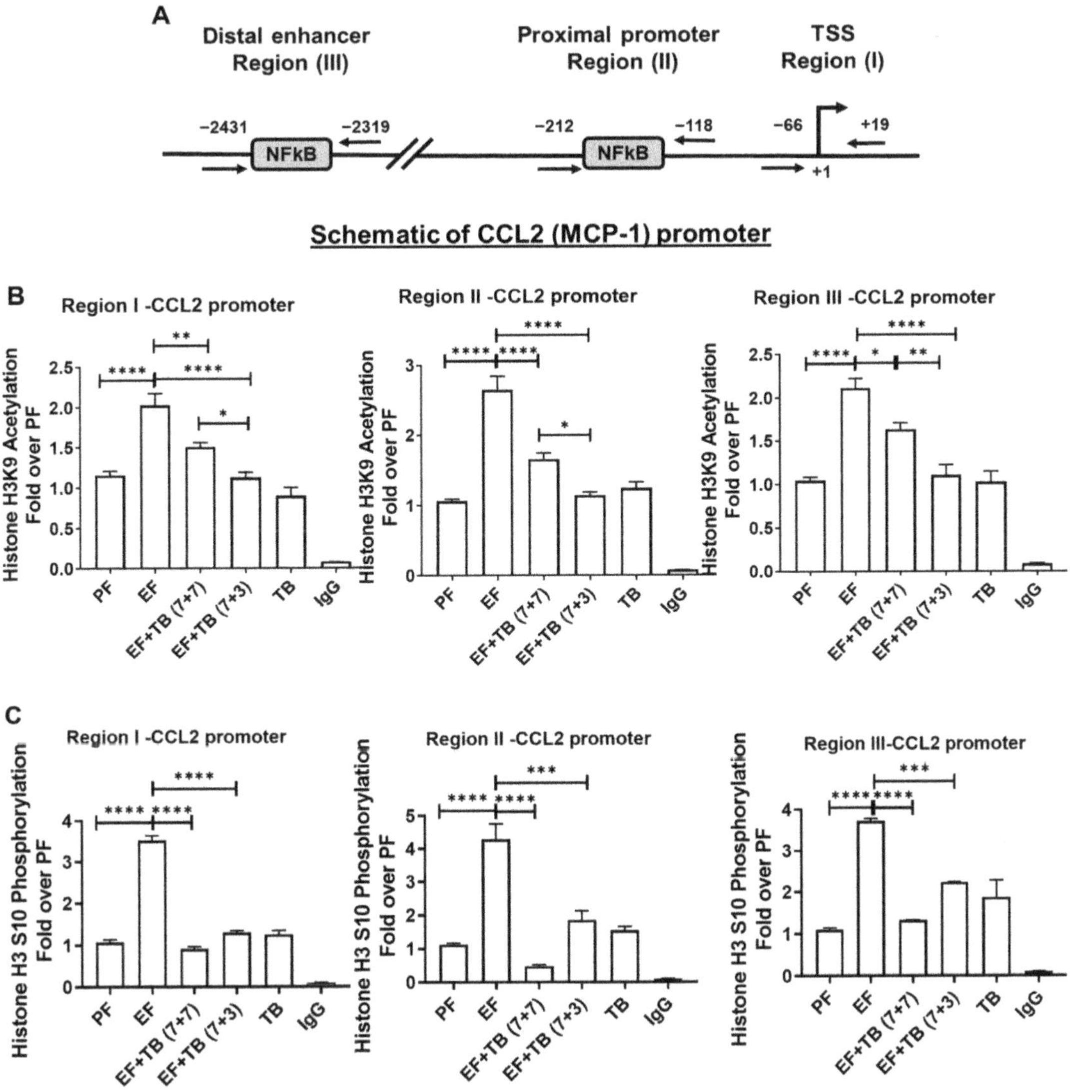

Figure 3. Effect of tributyrin administration on chronic ethanol-induced promoter-associated histone H3 modifications at CCL2 promoter in the liver: Mice were fed either a control diet (pair-fed, PF) or ethanol (ethanol-fed; EF) (5% *v*/*v*)-containing diet. Tributyrin (TB; 2 g/kg) was orally administered to ethanol-fed mice as a preventive (EF + TB (7 + 7)) or interventional (EF + TB (7 + 3)) treatment strategy. (**A**) Schematic of CCL2 promoter: Locations of transcription factor NFκB binding sites and ChIP-PCR primer pairs for analysis of epigenetic modifications are denoted as regions I–III. The coordinate locations shown are with respect to the transcription start site in REFSEQ NM_011333. Hepatic CCL2 promoter-associated histone modifications were assessed by analyzing chromatin that was immunoprecipitated with (**B**) acetylated anti–histone H3 lysine9 (H3K9) and (**C**) phosphorylated anti-histone H3 serine 10 (H3S10) antibodies. Levels of histone modifications were measured using primers specific for regions I, II, and III, as shown in the schematic. Differences are expressed as fold-over PF after normalizing for input DNA. Statistical analysis: mean ± SEM. * $p < 0.05$, ** $p < 0.01$, *** $p < 0.001$, and **** $p < 0.0001$ compared with PF or EF via ANOVA with Bonferroni's test (n = 5–8).

3.4. Tributyrin Impedes Ethanol-Responsive p300 Recruitment to the CCL2 Gene Promoter

Histone acetylation is regulated by histone acetyltransferases (HATs), and p300 is a known HAT that has been previously reported to increase histone acetylation at the CCL2 promoter [30,31]. Hence, we examined the changes in the p300 binding in response to ethanol and TB treatment. Similar to changes in histone H3K9 acetylation changes, ethanol feeding increased the p300 binding by ~7-fold at Region I, ~5-fold at Region II, and ~2.7-fold at Region III (Figure 4A) over PF. Interestingly, although being known as an HDAC activity inhibitor, tributyrin administration reduced p300 binding under both treatment strategies; however, the effect was more significant for interventional TB treatment (Figure 4A). There was no change in hepatic mRNA expression of p300 for any experimental groups. (Figure 4B).

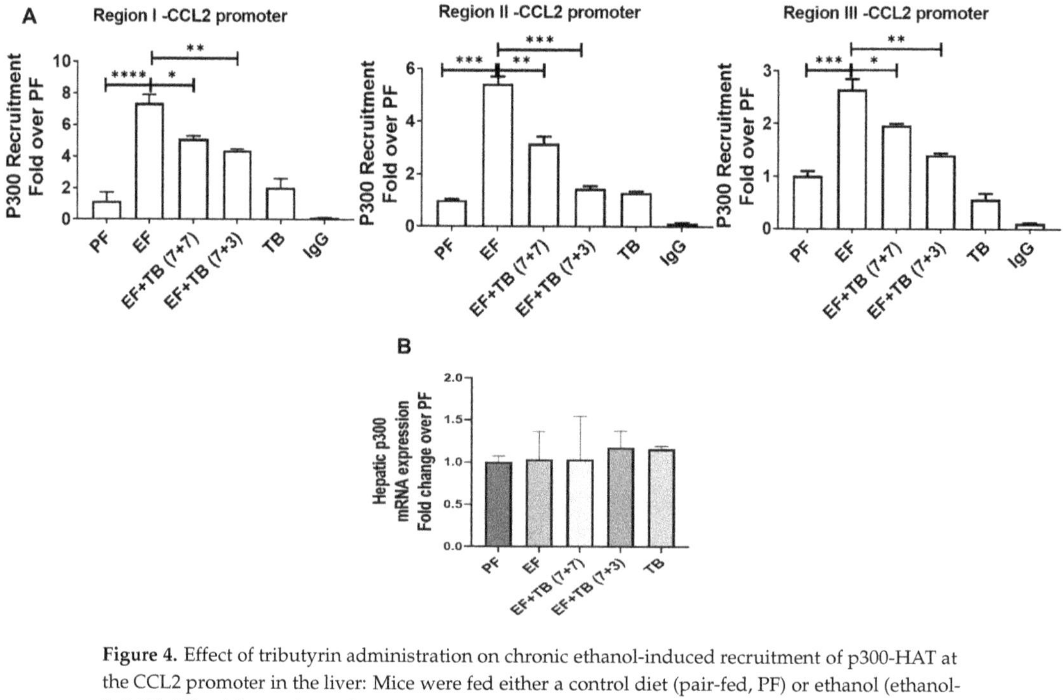

Figure 4. Effect of tributyrin administration on chronic ethanol-induced recruitment of p300-HAT at the CCL2 promoter in the liver: Mice were fed either a control diet (pair-fed, PF) or ethanol (ethanol-fed; EF) (5% *v*/*v*)-containing diet. Tributyrin (TB; 2 g/kg) was orally administered to ethanol-fed mice as a preventive (EF + TB (7 + 7)) or interventional (EF + TB (7 + 3)) treatment strategies. ChIP-qPCR quantification from (**A**) anti-p300 immunoprecipitated chromatin was performed. Differences are expressed as fold-over PF after normalizing for input DNA. (**B**) Hepatic mRNA analysis for p300 gene expression was performed using real-time PCR after normalizing with the TBP gene and expressed as a fold change over PF. Statistical analysis: mean ± SEM. * $p < 0.05$, ** $p < 0.01$, *** $p < 0.001$, and **** $p < 0.0001$ compared with PF or EF via ANOVA with Bonferroni's test (n = 5–8).

3.5. Tributyrin Attenuates Ethanol-Induced Enhanced Binding of Transcription Factor NFκB at the CCL2 Promoter

NFκB is one of the major transcription factors for CCL2 expression. Additionally, it is known that NFκB interacts with the co-activator protein p300 as part of the transcription initiation complex and recruits the complex to the target gene promoters [32]. Hence, we next examined the recruitment of NFκB to the CCL2 promoter via ChIP assay. Commensurate with increased p300 binding and upregulation of CCL2 mRNA expression, ethanol-exposed livers showed enhanced recruitment of NFκB at both proximal and distal enhancer regions of the CCL2 promoter. Moreover, TB treatment prevented NFκB binding to both proximal and distal enhancer regions of the CCL2 promoter (Figure 5A). There was no change in the mRNA expression of NFκB under these experimental conditions (Figure 5B).

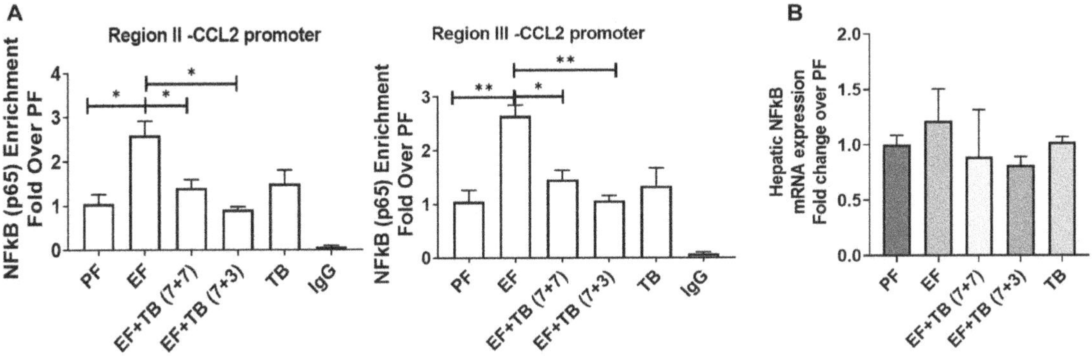

Figure 5. Effect of tributyrin administration on chronic ethanol-induced recruitment of the transcription factor NFκB at CCL2 promoter in the liver: Mice were fed either a control diet (pair-fed, PF) or ethanol (ethanol-fed; EF) (5% *v/v*)-containing diet. Tributyrin (TB; 2 g/kg) was orally administered to ethanol-fed mice as a preventive (EF + TB (7 + 7)) or interventional (EF + TB (7 + 3)) treatment strategy. (**A**) ChIP-qPCR quantification from immunoprecipitated chromatin using anti-NF-kB subunit RelA/p65 antibody was performed. Differences are expressed as a fold-over PF after normalizing for input DNA. (**B**) Hepatic mRNA analysis for NFκB was performed using real-time PCR after normalizing with the TBP gene and expressed as a fold change over PF. Statistical analysis: mean ± SEM. * $p < 0.05$, and ** $p < 0.01$ compared with PF or EF via ANOVA with Bonferroni's test (n = 5–8).

3.6. Tributyrin Modulates the SIRT1-NFκB-p300 Interaction and Transcriptional Activation of NFκB via Decreasing K310-Acetylation of the Nuclear RelA/p65 Subunit of NFκB in Ethanol-Fed Livers

Acetylation of lysine 310 (K310) of the p65 subunit of NFκB is required for its transcriptional activation and DNA-binding affinity [33,34]. Since increased NFκB occupancy was observed upon ethanol feeding, the acetylation status of lysine 310 of the nuclear p65 subunit was examined. Hepatic nuclear lysates were prepared, and Western blot analysis was performed to determine the levels of both total p65-NFκB and acetyl-K310-p65 (Figure 6A). There was a significant increase in the nuclear levels of p65 along with acetylation at K310 in the livers of mice fed ethanol. In contrast, the TB-administered groups showed a decrease in nuclear levels of p65 with a significant reduction in K310 acetylation of p65. These data suggest that TB suppressed the nuclear translocation and transcriptional activation of NFκB that occurred in response to ethanol feeding (Figure 6A).

Importantly, it has been reported that the nuclear acetylation status of K310 is maintained by the interaction of NFκB with a class III histone deacetylase—SIRT1—and histone acetyltransferase p300 [35–37]. Hence, the association of NFκB with SIRT1 and p300 was evaluated in ethanol-fed and tributyrin-administered mice. The p65 protein of NFκB was immunoprecipitated from hepatic nuclear lysates, and Western blot analysis was performed to determine the nuclear levels of Sirt1 and p300 associated with p65 (Figure 6B). We observed that chronic ethanol feeding modulated the SIRT1-NFκB-p300 axis in the liver. Specifically, the association of the nuclear RelA/p65 subunit of NFκB with a class III histone deacetylase, SIRT1, was decreased with a concomitant increase in a histone acetyltransferase, p300, interaction in livers of mice chronically fed ethanol compared to controls. On the other hand, TB treatment maintained the SIRT-1 association and reduced the p300 association with NFκB (Figure 6B). We also observed that, as previously shown, ethanol exposure decreased SIRT1 mRNA expression compared to the control. Interestingly, tributyrin, which is known as an HDAC inhibitor, increased hepatic mRNA expression of atypical HDAC-SIRT1 in ethanol-fed mice (Figure 6C).

3.7. TB Decreases the Binding of RNA Pol II in Ethanol-Fed Livers

In correlation with promoter histone hyper-acetylation and increased NFκB binding, ethanol enhanced the transcription initiation process by recruiting RNA Pol II at the TSS region I of the CCL2 promoter (Figure 7). Importantly, TB prevented these ethanol-induced permissive epigenetic effects and led to a decrease in RNA Pol II recruitment and CCL2 gene transcription (Figure 7).

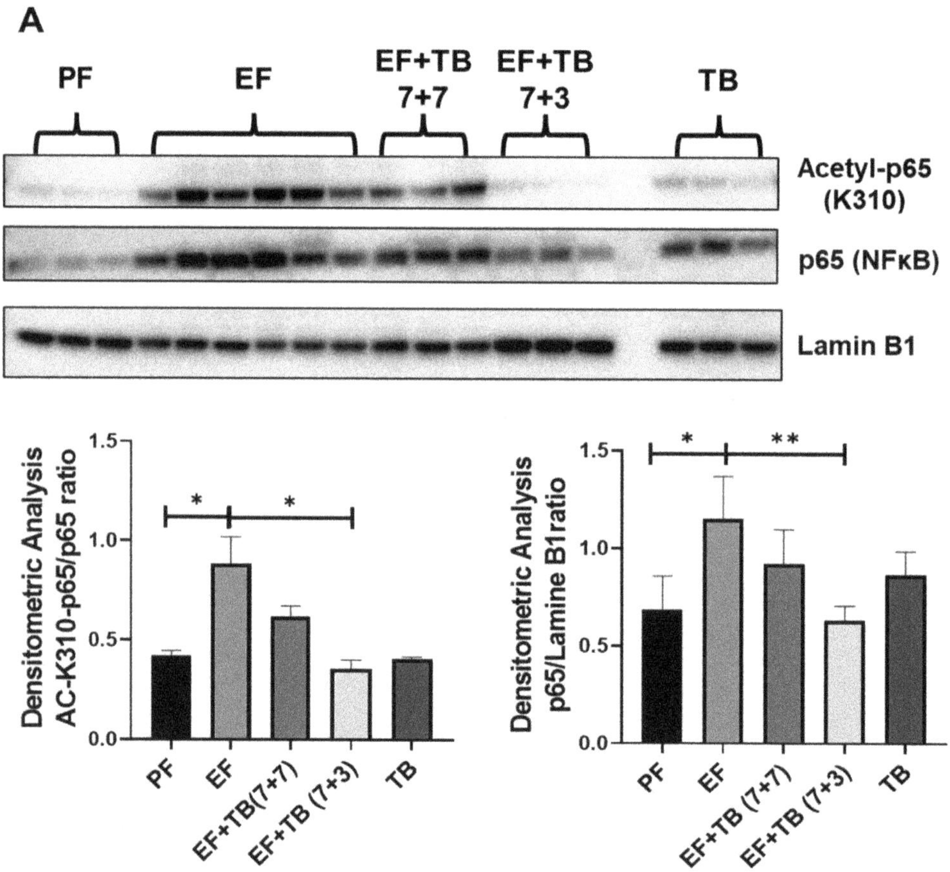

Figure 6. *Cont.*

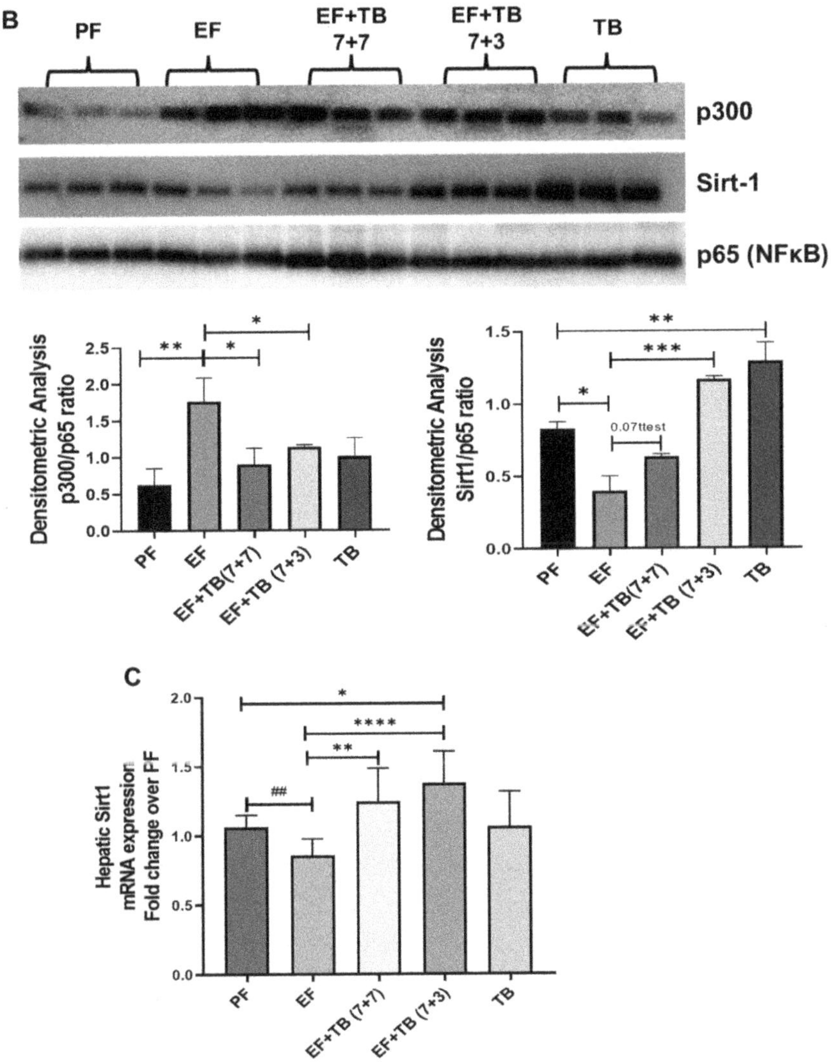

Figure 6. Effect of tributyrin administration on transcriptional activation of NFκB in chronic ethanol-fed mice livers: Mice were fed either a control diet (pair-fed, PF) or ethanol (ethanol-fed; EF) (5% *v*/*v*)-containing diet. Tributyrin (TB; 2 g/kg) was orally administered to ethanol-fed mice as a preventive (EF + TB (7 + 7)) or interventional (EF + TB (7 + 3)) treatment strategy. (**A**) Representative immunoblots for nuclear levels of lysine 310 acetylation (Acetyl-p65-K310) and total p65 subunit of NFκB were shown. The densitometric quantification for Acetyl-p65-K310 after normalizing to total p65 and levels of total p65 after normalizing to Lamine B are shown in the bar graphs. (**B**) Nuclear lysates from mice livers were immunoprecipitated with anti-NFκB-p65 antibodies and immunoblotted with anti-p300, SIRT-1, or NFκB-p65 antibodies. Representative immunoblots are shown. The densitometric quantification was performed, and p300 and SIRT-1 data normalized to NFκB-p65 are shown. (**C**) Hepatic mRNA analysis for SIRT1 was performed via real-time PCR and shown as fold change over PF. Data are presented as mean ± SEM, and a one-way-ANOVA analysis was used; significance is shown as * $p < 0.05$, ** $p < 0.01$, *** $p < 0.001$ and **** $p < 0.0001$. Analysis via Student's *t*-test is shown as ## $p < 0.01$.

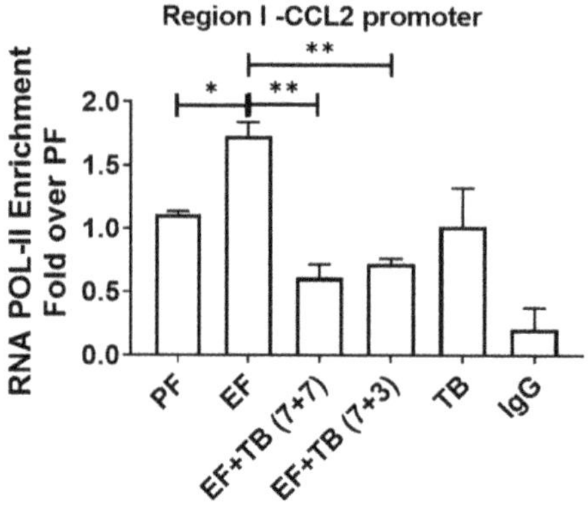

Figure 7. Effect of tributyrin administration on chronic ethanol-induced recruitment of RNA POL II at the CCL2 promoter in the liver: Mice were fed either a control diet (pair-fed, PF) or ethanol (ethanol-fed; EF) (5% *v/v*)-containing diet. Tributyrin (TB; 2 g/kg) was orally administered to ethanol-fed mice as a preventive (EF + TB (7 + 7)) or interventional (EF + TB (7 + 3)) treatment strategy. ChIP-qPCR quantification from immunoprecipitated chromatin using an anti-RNA POL II antibody was performed. Differences are expressed as fold-over PF after normalizing for input DNA. Statistical analysis: mean ± SEM. * $p < 0.05$, and ** $p < 0.01$ compared with PF or EF via ANOVA with Bonferroni's test (n = 5–8).

Taken together, our study indicates that chronic ethanol-induced hyperacetylation of histone H3(K9) and the transcription factor NFκB (K310) drive hepatic CCL2 expression and, in turn, contribute to liver inflammation. Importantly, transcriptional repression of CCL2 via tributyrin administration, under both preventive and interventional treatment strategies, was associated with hypoacetylation and diminished binding of chromatin-associated proteins, including p300, NFκB, and RNA polymerase II.

4. Discussion

Alcohol-associated liver disease (ALD) is a major cause of liver-related morbidity and mortality [1,38,39]. Despite being actively investigated, there is no FDA-approved therapy for any stage of ALD. The development of ALD is a complex process that is influenced by a variety of genetic and environmental factors. Research over the past decade has determined that combined effects of alcohol metabolism and compromised nutritional status are associated with disease pathogenesis involving alterations in gut microbiota, key metabolites such as SCFAs, barrier dysfunction, and liver inflammation and injury [18]. In the context of liver inflammation, both clinical and animal studies of ALD have provided evidence that hepatic expression of CCL2 was correlated with the recruitment of macrophages and neutrophils and the severity of liver inflammation and injury [5,7]; however, the mechanisms underlying their upregulation are not completely elucidated. The Ccl2 gene is transcribed at low levels under noninflammatory conditions, and exposure to various pro-inflammatory stimuli/conditions leads to the rapid induction of Ccl2 gene expression [28]. Since the differential Ccl2 gene expression is controlled via an epigenetic mechanism, the present study examined the key aspects of epigenetic regulation of ethanol-inducible CCL2 expression employing the chronic ethanol-feeding mouse model of ALD. The major finding of the present study demonstrated that alterations in the post-translational acetylation modification of both histone H3 and transcription factor NFκB by p300 and SIRT1 play an important role in regulating hepatic CCL2 expression in response to both ethanol and oral administration of tributyrin.

Under both normal and disease states, the epigenetic machinery is important for regulating gene expression. Post-translational modifications (PTMs) of proteins are considered the main components of epigenetic regulatory machinery for gene expression and are also broadly accepted as a therapeutic target [40]. Various post-translational modifications of histone H3 can occur, and depending on the type of modification, it creates a euchromatic or heterochromatic chromatin structure at the promoter, thereby affecting gene expression [41]. Histone H3 acetylation always leads to an open chromatin structure supporting transcription factor binding and enhanced gene transcription. In our study, we observed that ethanol feeding increased H3K9Ac levels at the CCL2 promoter, leading to transcriptional activation and CCL2 gene upregulation. This result was in line with previously published studies where an increase in histone H3K9 acetylation at CCL2 has been reported in obesity and fatty liver [42]. Moreover, ethanol exposure is known to increase global H3K9 acetylation levels in rat hepatocytes [43]. Another documented permissive modification of histone H3 is phosphorylation at serine 10. It is known that H3S10ph not only affects H3K9 acetylation but also acts in synergy with it and increases the efficiency of acetylation reactions [44,45]. Studies by Park et al. have shown a reduction in ethanol-induced acetylation after inhibiting the kinases that are known to phosphorylate Histone H3 at serine 10, suggesting a crosstalk between these two adjacent modifications [43]. We also observed a concomitant change in H3K9 acetylation and H3S10 phosphorylation at the CCL2 promoter under ethanol and tributyrin feeding conditions, further supporting their interactive role in regulating gene transcription. Additionally, ethanol-mediated increases in H3K9 acetylation and H3S10 phosphorylation also enhanced the DNA accessibility for transcriptional machinery, as seen by more binding of NFκB and RNA PoL II upregulating CCL2 transcription.

In addition to histones, acetylation of several non-histone proteins, including transcription factors, also plays an important role in regulating their function and, in turn, target gene transcription [46,47]. NFκB is one such transcription factor whose function is affected by the acetylation of lysine residues [33,48]. Importantly, NFκB is a predominant transcription factor regulating CCL2 gene expression [28,29,49]. Among the seven lysine residues (lysine 122, 123, 218, 221, 310, 314, and 315) of NFκB that are known to become acetylated, acetylation of lysine 310 is important because it is required for full activation of NF-kB [34]. Moreover, it has been reported that lysine 310 is acetylated by HAT-p300 [35], and mutation of this lysine residue does not allow functional cooperation of p65 with p300/CBP and thus impairs the transactivation of p65 [34]. In this context, we observed that livers from ethanol-fed mice showed greater interaction between the p300 and p65 subunit of NFκB and an increase in K310 acetylation of p65.

Thus, these data suggest that ethanol exposure enhanced the transcription potential of NFκB, increasing its binding to the CCL2 promoter and consequent CCL2 upregulation. Additionally, it is known that the interaction of NFκB with co-activator proteins, such as p300, allows their recruitment to target gene promoters [50,51]. The p300, being a histone acetyltransferase (HAT), also acetylates the target gene promoters, creating an open chromatin structure that is permissive for gene transcription. Our data also showed a significant increase in p300 binding to the CCL2 promoter, which then caused the hyperacetylation of histone H3K9 in response to ethanol. These observations suggest that under ethanol conditions, p300 mediates both acetylating histone H3 at lysine 9 and NFκB at lysine 310, thus enhancing the transcription of hepatic CCL2 expression. It is known that alcohol metabolism leads to an increase in acetate levels and acetyl COA levels, which is the main substrate for the p300 acetylation function. Although we have not evaluated the levels of acetyl CoA in our study, based on our results, it can be postulated that ethanol metabolism increases the substrate availability of p300 to acetylate both histone H3 and transcription factor NFκB-p65.

In contrast to ethanol, both preventive and interventional treatment strategies of tributyrin administration showed that TB treatment inhibited the ethanol-inducible CCL2 expression. Tributyrin is a triglyceride containing three butyrate moieties that gets rapidly absorbed and hydrolyzed to butyrate [15,52]. We have previously shown that oral administrations of tributyrin prevent alcohol-induced microbial dysbiosis, mainly loss of butyrate-producing bacteria, and can increase hepato-portal circulation of the butyrate, leading to a direct effect on the liver and protect against steatohepatitis in mice [16]. In the context of inflammation, TB has been shown to reduce the expression of pro-inflammatory cytokines, including MCP-1/CCL2, in adipose tissue in high-fat-fed mice [53]. In the present study, we showed that the protective effects of TB are through decreasing ethanol-mediated permissive acetylation modifications of promoter-associated histones and the p65 subunit of NFκB. Although TB is a dietary butyrate prodrug and has been documented to have HDAC inhibitory function, our study demonstrated that TB increases the hepatic expression and functional capacity of atypical HDAC-SIRT1. Sirtuin 1(SIRT1) is a nicotinamide adenine dinucleotide (NAD+, NADH)-dependent class III histone deacetylase, and its role in alcohol-associated fatty liver disease has been well documented [54]. It has been reported that ethanol downregulates SIRT1 expression in the liver, and ethanol-mediated disruption of SIRT1 signaling leads to fat accumulation and inflammatory responses in the liver [55]. In agreement with the earlier studies, our data also demonstrated a significant decrease in SIRT1 expression in ethanol-fed animals; however, TB treatment not only restored the ethanol-induced downregulation of SIRT1 mRNA expression but further increased its expression. Consistent with the documented results, butyrate has been shown to increase SIRT1 expression in the colon and alleviate DSS-induced inflammation in mice [56]. Moreover, relevant to the present work, it has been reported that changes in p65 acetylation status are regulated by the interaction between p65 and SIRT1 and deacetylation by SIRT1 occurs at lysine 310 [37]. Indeed, our data showed that commensurate with its expression, TB also increases the functional capacity of SIRT1 to interact with and deacetylate the p65 subunit of NFκB. Our data showed that the SIRT1 acts as a negative regulator of NFκB function in TB-treated livers as seen by an increased nuclear association between SIRT-1 and p65 with a significant decrease in acetylation at K310 and subsequent decreased recruitment to CCL2 promoter. The present data are consistent with the previous studies documenting that the use of pharmacological agents modulating SIRT1 activity affects the acetylation status of RelA protein at Lys310 and its inflammatory transactivation potential [35]. It was shown in a SIRT1 knockout mouse model that the deletion of SIRT1, hyperacetylated NFκB and increases its activity, thus resulting in the upregulation of pro-inflammatory genes [57].

TB treatment also interferes with p65 binding to p300, which occurs in response to ethanol exposure. This resultant decrease in p300-p65 association likely contributed to further deacetylation of NFκB and reduced recruitment of p300 to CCL2 promoter, decreasing the H3K9 acetylation levels. Thus, oral administration of tributyrin can induce close chromatin structure and transcriptional suppression of CCL2 expression. It has been reported that SIRT1 can physically interact with p300, inactivating its acetyltransferase capacity [58,59]. Hence, it can be easily postulated that TB exerts its protective effect mainly via modulating HAT-p300 and typical HDAC-SIRT-1 interaction with the p65 subunit of NFκB and deacetylating p65 and impairing its recruitment to CCL2 promoter. Since NFκB is a key regulator of several inflammatory processes, downregulation of NFκB transcriptional activation via tributyrin implies that targeting the SIRT1-NFκB-p300 association may provide new therapeutic opportunities not only for the treatment of ALD but also for other inflammatory conditions. However, more research is needed to fully understand the role of these interactions in ALD and determine if targeting them may be an effective strategy for preventing the development and progression of ALD.

5. Conclusions

Overall, our data support the idea that not only acetylation of histone H3 at the CCL2 promoter but also acetylation and transactivation of the non-histone protein, NFκB-p65, contribute to alcohol-induced upregulation of CCL2 in the liver.

Author Contributions: Study design and concept: S.S.G. and S.S.B.; performed experiments: S.S.G., D.T.G., B.T.C., M.V.V. and S.R.; animal histology analysis: J.Z.; data analysis and interpretation: S.S.G., S.J.-B., C.J.M. and S.S.B.; manuscript preparation: S.S.G., C.J.M. and S.S.B. All authors have read and agreed to the published version of the manuscript.

Funding: Research reported in this publication was supported by the National Institute Of General Medical Sciences of the National Institutes of Health under Award Number P20GM113226 (CJM/SSB/SSG); the National Institute On Alcohol Abuse and Alcoholism of the National Institutes of Health under Award Number P50AA024337 (CJM/SSB); RO1AA024405 (SSB); PO1 AA029542(SSB); PO1AA029543(SSB); UO1AA022618 (SSB); U01AA026936 (CJM/SJB); 1U01AA026934 (CJM); 1U01AA026980 (CJM); and the VA (CJM). The content is solely the responsibility of the authors and does not necessarily represent the official views of the National Institutes of Health.

Institutional Review Board Statement: All experiments were carried out according to the criteria outlined in the Guide for Care and Use of Laboratory Animals and with the approval of the University of Louisville's Institutional Animal Care and Use Committee (IACUC) number 20754 on 20 August 2020.

Data Availability Statement: Data will be made available upon reasonable request.

Conflicts of Interest: The authors declare no conflict of interest.

Abbreviations

ALD, alcohol associated liver disease; MCP-1, monocyte chemoattractant protein-1; CCL2, C-C motif chemokine ligand 2; SCFA, short-chain fatty acid; ChIP, chromatin immunoprecipitation; HAT, histone acetyltransferase; HDAC, histone deacetylase; H3K9Ac, histone 3 lysine 9 acetylation; qPCR, quantitative real-time PCR; RNA Pol II, RNA polymerase II; TSS, transcription start site.

References

1. Bruha, R.; Dvorak, K.; Petrtyl, J. Alcoholic liver disease. *World J. Hepatol.* **2012**, *4*, 81–90. [CrossRef] [PubMed]
2. Bautista, A.P. Neutrophilic infiltration in alcoholic hepatitis. *Alcohol* **2002**, *27*, 17–21. [CrossRef] [PubMed]
3. Gao, B.; Xu, M. Chemokines and alcoholic hepatitis: Are chemokines good therapeutic targets? *Gut* **2014**, *63*, 1683–1684. [CrossRef] [PubMed]
4. Wang, H.J.; Gao, B.; Zakhari, S.; Nagy, L.E. Inflammation in alcoholic liver disease. *Annu. Rev. Nutr.* **2012**, *32*, 343–368. [CrossRef] [PubMed]
5. Degre, D.; Lemmers, A.; Gustot, T.; Ouziel, R.; Trepo, E.; Demetter, P.; Verset, L.; Quertinmont, E.; Vercruysse, V.; Le Moine, O.; et al. Hepatic expression of CCL2 in alcoholic liver disease is associated with disease severity and neutrophil infiltrates. *Clin. Exp. Immunol.* **2012**, *169*, 302–310. [CrossRef] [PubMed]
6. Hill, D.B.; Marsano, L.S.; McClain, C.J. Increased plasma interleukin-8 concentrations in alcoholic hepatitis. *Hepatology* **1993**, *18*, 576–580. [CrossRef]
7. Mandrekar, P.; Ambade, A.; Lim, A.; Szabo, G.; Catalano, D. An essential role for monocyte chemoattractant protein-1 in alcoholic liver injury: Regulation of proinflammatory cytokines and hepatic steatosis in mice. *Hepatology* **2011**, *54*, 2185–2197. [CrossRef] [PubMed]
8. Xie, G.; Zhong, W.; Zheng, X.; Li, Q.; Qiu, Y.; Li, H.; Chen, H.; Zhou, Z.; Jia, W. Chronic ethanol consumption alters mammalian gastrointestinal content metabolites. *J. Proteome Res.* **2013**, *12*, 3297–3306. [CrossRef]
9. Berni Canani, R.; Di Costanzo, M.; Leone, L. The epigenetic effects of butyrate: Potential therapeutic implications for clinical practice. *Clin. Epigenet.* **2012**, *4*, 4. [CrossRef]
10. Mathew, O.P.; Ranganna, K.; Yatsu, F.M. Butyrate, an HDAC inhibitor, stimulates interplay between different posttranslational modifications of histone H3 and differently alters G1-specific cell cycle proteins in vascular smooth muscle cells. *Biomed. Pharmacother.* **2010**, *64*, 733–740. [CrossRef]
11. Mattis, E.R.; Ostapiuk, S.F. The trends in the management reorganization of medical science under market economy conditions. *Vestn. Ross. Akad. Meditsinskikh Nauk.* **1992**, *11–12*, 61–63.
12. Segain, J.P.; Raingeard de la Bletiere, D.; Bourreille, A.; Leray, V.; Gervois, N.; Rosales, C.; Ferrier, L.; Bonnet, C.; Blottiere, H.M.; Galmiche, J.P. Butyrate inhibits inflammatory responses through NFkappaB inhibition: Implications for Crohn's disease. *Gut* **2000**, *47*, 397–403. [CrossRef] [PubMed]
13. Inan, M.S.; Rasoulpour, R.J.; Yin, L.; Hubbard, A.K.; Rosenberg, D.W.; Giardina, C. The luminal short-chain fatty acid butyrate modulates NF-kappaB activity in a human colonic epithelial cell line. *Gastroenterology* **2000**, *118*, 724–734. [CrossRef] [PubMed]
14. Schreiber, S.; Nikolaus, S.; Hampe, J. Activation of nuclear factor kappa B inflammatory bowel disease. *Gut* **1998**, *42*, 477–484. [CrossRef] [PubMed]
15. Cresci, G.A.; Glueck, B.; McMullen, M.R.; Xin, W.; Allende, D.; Nagy, L.E. Prophylactic tributyrin treatment mitigates chronic-binge ethanol-induced intestinal barrier and liver injury. *J. Gastroenterol. Hepatol.* **2017**, *32*, 1587–1597. [CrossRef] [PubMed]

16. Donde, H.; Ghare, S.; Joshi-Barve, S.; Zhang, J.; Vadhanam, M.V.; Gobejishvili, L.; Lorkiewicz, P.; Srivastava, S.; McClain, C.J.; Barve, S. Tributyrin Inhibits Ethanol-Induced Epigenetic Repression of CPT-1A and Attenuates Hepatic Steatosis and Injury. *Cell Mol. Gastroenterol. Hepatol.* **2020**, *9*, 569–585. [CrossRef] [PubMed]
17. Shvedunova, M.; Akhtar, A. Modulation of cellular processes by histone and non-histone protein acetylation. *Nat. Rev. Mol. Cell Biol.* **2022**, *23*, 329–349. [CrossRef] [PubMed]
18. Moghe, A.; Joshi-Barve, S.; Ghare, S.; Gobejishvili, L.; Kirpich, I.; McClain, C.J.; Barve, S. Histone modifications and alcohol-induced liver disease: Are altered nutrients the missing link? *World J. Gastroenterol.* **2011**, *17*, 2465–2472. [CrossRef]
19. Eberharter, A.; Becker, P.B. Histone acetylation: A switch between repressive and permissive chromatin. Second in review series on chromatin dynamics. *EMBO Rep.* **2002**, *3*, 224–229. [CrossRef]
20. Kirpich, I.; Ghare, S.; Zhang, J.; Gobejishvili, L.; Kharebava, G.; Barve, S.J.; Barker, D.; Moghe, A.; McClain, C.J.; Barve, S. Binge alcohol-induced microvesicular liver steatosis and injury are associated with down-regulation of hepatic Hdac 1, 7, 9, 10, 11 and up-regulation of Hdac 3. *Alcohol. Clin. Exp. Res.* **2012**, *36*, 1578–1586. [CrossRef]
21. Osna, N.A.; Carter, W.G.; Ganesan, M.; Kirpich, I.A.; McClain, C.J.; Petersen, D.R.; Shearn, C.T.; Tomasi, M.L.; Kharbanda, K.K. Aberrant post-translational protein modifications in the pathogenesis of alcohol-induced liver injury. *World J. Gastroenterol.* **2016**, *22*, 6192–6200. [CrossRef] [PubMed]
22. Park, P.H.; Lim, R.W.; Shukla, S.D. Gene-selective histone H3 acetylation in the absence of increase in global histone acetylation in liver of rats chronically fed alcohol. *Alcohol Alcohol.* **2012**, *47*, 233–239. [CrossRef] [PubMed]
23. Kim, J.S.; Shukla, S.D. Acute in vivo effect of ethanol (binge drinking) on histone H3 modifications in rat tissues. *Alcohol Alcohol.* **2006**, *41*, 126–132. [CrossRef] [PubMed]
24. Park, P.H.; Miller, R.; Shukla, S.D. Acetylation of histone H3 at lysine 9 by ethanol in rat hepatocytes. *Biochem. Biophys. Res. Commun.* **2003**, *306*, 501–504. [CrossRef] [PubMed]
25. Ghare, S.S.; Joshi-Barve, S.; Moghe, A.; Patil, M.; Barker, D.F.; Gobejishvili, L.; Brock, G.N.; Cave, M.; McClain, C.J.; Barve, S.S. Coordinated histone H3 methylation and acetylation regulate physiologic and pathologic fas ligand gene expression in human CD4+ T cells. *J. Immunol.* **2014**, *193*, 412–421. [CrossRef] [PubMed]
26. Ghare, S.; Patil, M.; Hote, P.; Suttles, J.; McClain, C.; Barve, S.; Joshi-Barve, S. Ethanol inhibits lipid raft-mediated TCR signaling and IL-2 expression: Potential mechanism of alcohol-induced immune suppression. *Alcohol. Clin. Exp. Res.* **2011**, *35*, 1435–1444. [CrossRef] [PubMed]
27. Deng, X.; Xu, M.; Yuan, C.; Yin, L.; Chen, X.; Zhou, X.; Li, G.; Fu, Y.; Feghali-Bostwick, C.A.; Pang, L. Transcriptional regulation of increased CCL2 expression in pulmonary fibrosis involves nuclear factor-kappaB and activator protein-1. *Int. J. Biochem. Cell Biol.* **2013**, *45*, 1366–1376. [CrossRef]
28. Hildebrand, D.G.; Alexander, E.; Horber, S.; Lehle, S.; Obermayer, K.; Munck, N.A.; Rothfuss, O.; Frick, J.S.; Morimatsu, M.; Schmitz, I.; et al. IkappaBzeta is a transcriptional key regulator of CCL2/MCP-1. *J. Immunol.* **2013**, *190*, 4812–4820. [CrossRef]
29. Ueda, A.; Ishigatsubo, Y.; Okubo, T.; Yoshimura, T. Transcriptional regulation of the human monocyte chemoattractant protein-1 gene. Cooperation of two NF-kappaB sites and NF-kappaB/Rel subunit specificity. *J. Biol. Chem.* **1997**, *272*, 31092–31099. [CrossRef]
30. Gao, J.; Wei, B.; Liu, M.; Hirsova, P.; Sehrawat, T.S.; Cao, S.; Hu, X.; Xue, F.; Yaqoob, U.; Kang, N.; et al. Endothelial p300 Promotes Portal Hypertension and Hepatic Fibrosis Through C-C Motif Chemokine Ligand 2-Mediated Angiocrine Signaling. *Hepatology* **2021**, *73*, 2468–2483. [CrossRef]
31. Deng, X.; Zhou, X.; Deng, Y.; Liu, F.; Feng, X.; Yin, Q.; Gu, Y.; Shi, S.; Xu, M. Thrombin Induces CCL2 Expression in Human Lung Fibroblasts via p300 Mediated Histone Acetylation and NF-KappaB Activation. *J. Cell Biochem.* **2017**, *118*, 4012–4019. [CrossRef] [PubMed]
32. Mukherjee, S.P.; Behar, M.; Birnbaum, H.A.; Hoffmann, A.; Wright, P.E.; Ghosh, G. Analysis of the RelA:CBP/p300 interaction reveals its involvement in NF-kappaB-driven transcription. *PLoS Biol.* **2013**, *11*, e1001647. [CrossRef] [PubMed]
33. Giridharan, S.; Srinivasan, M. Mechanisms of NF-kappaB p65 and strategies for therapeutic manipulation. *J. Inflamm. Res.* **2018**, *11*, 407–419. [CrossRef] [PubMed]
34. Chen, L.F.; Mu, Y.; Greene, W.C. Acetylation of RelA at discrete sites regulates distinct nuclear functions of NF-kappaB. *EMBO J.* **2002**, *21*, 6539–6548. [CrossRef] [PubMed]
35. Yang, H.; Zhang, W.; Pan, H.; Feldser, H.G.; Lainez, E.; Miller, C.; Leung, S.; Zhong, Z.; Zhao, H.; Sweitzer, S.; et al. SIRT1 activators suppress inflammatory responses through promotion of p65 deacetylation and inhibition of NF-kappaB activity. *PLoS ONE* **2012**, *7*, e46364. [CrossRef]
36. Huang, B.; Yang, X.D.; Lamb, A.; Chen, L.F. Posttranslational modifications of NF-kappaB: Another layer of regulation for NF-kappaB signaling pathway. *Cell Signal* **2010**, *22*, 1282–1290. [CrossRef] [PubMed]
37. Yeung, F.; Hoberg, J.E.; Ramsey, C.S.; Keller, M.D.; Jones, D.R.; Frye, R.A.; Mayo, M.W. Modulation of NF-kappaB-dependent transcription and cell survival by the SIRT1 deacetylase. *EMBO J* **2004**, *23*, 2369–2380. [CrossRef]
38. Chacko, K.R.; Reinus, J. Spectrum of Alcoholic Liver Disease. *Clin. Liver. Dis.* **2016**, *20*, 419–427. [CrossRef]
39. Farooq, M.O.; Bataller, R. Pathogenesis and Management of Alcoholic Liver Disease. *Dig. Dis.* **2016**, *34*, 347–355. [CrossRef]
40. Siklos, M.; Kubicek, S. Therapeutic targeting of chromatin: Status and opportunities. *FEBS J.* **2022**, *289*, 1276–1301. [CrossRef]
41. Swygert, S.G.; Peterson, C.L. Chromatin dynamics: Interplay between remodeling enzymes and histone modifications. *Biochim. Biophys. Acta* **2014**, *1839*, 728–736. [CrossRef] [PubMed]

42. Mikula, M.; Majewska, A.; Ledwon, J.K.; Dzwonek, A.; Ostrowski, J. Obesity increases histone H3 lysine 9 and 18 acetylation at Tnfa and Ccl2 genes in mouse liver. *Int. J. Mol. Med.* **2014**, *34*, 1647–1654. [CrossRef] [PubMed]
43. Park, P.H.; Lim, R.W.; Shukla, S.D. Involvement of histone acetyltransferase (HAT) in ethanol-induced acetylation of histone H3 in hepatocytes: Potential mechanism for gene expression. *Am. J. Physiol. Gastrointest. Liver. Physiol.* **2005**, *289*, G1124–G1136. [CrossRef] [PubMed]
44. Keppler, B.R.; Archer, T.K. Chromatin-modifying enzymes as therapeutic targets—Part 2. *Expert Opin. Ther. Targets* **2008**, *12*, 1457–1467. [CrossRef] [PubMed]
45. Zhong, S.; Goto, H.; Inagaki, M.; Dong, Z. Phosphorylation at serine 28 and acetylation at lysine 9 of histone H3 induced by trichostatin A. *Oncogene* **2003**, *22*, 5291–5297. [CrossRef] [PubMed]
46. Xia, C.; Tao, Y.; Li, M.; Che, T.; Qu, J. Protein acetylation and deacetylation: An important regulatory modification in gene transcription (Review). *Exp. Ther. Med.* **2020**, *20*, 2923–2940. [CrossRef] [PubMed]
47. Park, J.M.; Jo, S.H.; Kim, M.Y.; Kim, T.H.; Ahn, Y.H. Role of transcription factor acetylation in the regulation of metabolic homeostasis. *Protein Cell* **2015**, *6*, 804–813. [CrossRef]
48. Ghizzoni, M.; Haisma, H.J.; Maarsingh, H.; Dekker, F.J. Histone acetyltransferases are crucial regulators in NF-kappaB mediated inflammation. *Drug Discov. Today* **2011**, *16*, 504–511. [CrossRef]
49. Wang, Y.; Rangan, G.K.; Goodwin, B.; Tay, Y.C.; Harris, D.C. Lipopolysaccharide-induced MCP-1 gene expression in rat tubular epithelial cells is nuclear factor-kappaB dependent. *Kidney Int.* **2000**, *57*, 2011–2022. [CrossRef]
50. Chan, H.M.; La Thangue, N.B. p300/CBP proteins: HATs for transcriptional bridges and scaffolds. *J. Cell Sci.* **2001**, *114*, 2363–2373. [CrossRef]
51. Ogryzko, V.V.; Schiltz, R.L.; Russanova, V.; Howard, B.H.; Nakatani, Y. The transcriptional coactivators p300 and CBP are histone acetyltransferases. *Cell* **1996**, *87*, 953–959. [CrossRef] [PubMed]
52. Miyoshi, M.; Sakaki, H.; Usami, M.; Iizuka, N.; Shuno, K.; Aoyama, M.; Usami, Y. Oral administration of tributyrin increases concentration of butyrate in the portal vein and prevents lipopolysaccharide-induced liver injury in rats. *Clin. Nutr.* **2011**, *30*, 252–258. [CrossRef] [PubMed]
53. Vinolo, M.A.; Rodrigues, H.G.; Festuccia, W.T.; Crisma, A.R.; Alves, V.S.; Martins, A.R.; Amaral, C.L.; Fiamoncini, J.; Hirabara, S.M.; Sato, F.T.; et al. Tributyrin attenuates obesity-associated inflammation and insulin resistance in high-fat-fed mice. *Am. J. Physiol. Endocrinol. Metab.* **2012**, *303*, E272–E282. [CrossRef] [PubMed]
54. Ren, R.; Wang, Z.; Wu, M.; Wang, H. Emerging Roles of SIRT1 in Alcoholic Liver Disease. *Int. J. Biol. Sci.* **2020**, *16*, 3174–3183. [CrossRef]
55. You, M.; Jogasuria, A.; Taylor, C.; Wu, J. Sirtuin 1 signaling and alcoholic fatty liver disease. *Hepatobiliary Surg. Nutr.* **2015**, *4*, 88–100. [CrossRef]
56. Wang, S.; Deng, W.; Li, F.; Xiang, L.; Lv, P.; Chen, Y. Treatment with butyrate alleviates dextran sulfate sodium and Clostridium difficile-induced colitis by preventing activity of Th17 cells via regulation of SIRT1/mTOR in mice. *J. Nutr. Biochem.* **2023**, *111*, 109155. [CrossRef]
57. Schug, T.T.; Xu, Q.; Gao, H.; Peres-da-Silva, A.; Draper, D.W.; Fessler, M.B.; Purushotham, A.; Li, X. Myeloid deletion of SIRT1 induces inflammatory signaling in response to environmental stress. *Mol. Cell Biol.* **2010**, *30*, 4712–4721. [CrossRef]
58. Shakibaei, M.; Buhrmann, C.; Mobasheri, A. Resveratrol-mediated SIRT-1 interactions with p300 modulate receptor activator of NF-kappaB ligand (RANKL) activation of NF-kappaB signaling and inhibit osteoclastogenesis in bone-derived cells. *J. Biol. Chem.* **2011**, *286*, 11492–11505. [CrossRef]
59. Bouras, T.; Fu, M.; Sauve, A.A.; Wang, F.; Quong, A.A.; Perkins, N.D.; Hay, R.T.; Gu, W.; Pestell, R.G. SIRT1 deacetylation and repression of p300 involves lysine residues 1020/1024 within the cell cycle regulatory domain 1. *J. Biol. Chem.* **2005**, *280*, 10264–10276. [CrossRef]

nutrients

MDPI

Article

Interactions between Polygenetic Variants and Lifestyle Factors in Hypothyroidism: A Hospital-Based Cohort Study

Da Sol Kim and Sunmin Park *

Department of Food and Nutrition, Obesity/Diabetes Research Center, Hoseo University, Asan 31499, Republic of Korea; tpfptm14@daum.net
* Correspondence: smpark@hoseo.edu; Tel.: +82-41-540-5345; Fax:+82-41-548-0670

Abstract: Hypothyroidism is a prevalent endocrine disorder and is associated with a variety of metabolic disturbances. This study aimed to investigate the polygenic variants associated with hypothyroidism risk and the interaction of polygenic risk scores (PRS) with dietary patterns in influencing disease risk in 56,664 participants aged >40 in a hospital-based cohort. The participants were classified as having hypothyroidism (n = 870) diagnosed by a physician and no hypothyroidism (n = 55,794). Genetic variants associated with hypothyroidism were identified using a genome-wide association study (GWAS). Genetic variants interacting with each other were selected using a generalized multifactor dimensionality reduction analysis, and the PRS generated was evaluated for interaction with lifestyle parameters. Coffee, alcohol, meat intake, and a Korean balanced diet were inversely associated with hypothyroidism risk, as were selenium, copper, and manganese intakes. White blood cell (WBC) counts and serum alkaline phosphatase and triglyceride concentrations were positively associated with hypothyroidism risk, as were osteoporosis and thyroid cancer. The GMDR analysis generated a three-single nucleotide polymorphism (SNP) model comprising dual oxidase-1 (*DUOX1*)_rs1648314; thyroid-stimulating hormone receptor (*TSHR*)_rs75664963; and major histocompatibility complex, class-II, DQ Alpha-1 (*HLA-DQA1*)_rs17426593. The PRS derived from the three- and seven-SNP models were associated with a 2.11- and 2.32-fold increase in hypothyroidism risk, respectively. Furthermore, the PRS from the three-SNP model showed interactions with WBC counts, wherein the positive association with hypothyroidism risk was more pronounced in participants with low WBC counts than those with high WBC counts ($\geq 4 \times 10^9$ /L). Dietary patterns, such as the plant-based diet (PBD) and the Western-style diet (WSD), along with smoking status, exhibited interactions with the PRS, influencing hypothyroidism risk. In participants with a high PRS, those in the high-PBD, low-WSD, and smoker groups had a higher proportion of hypothyroidism than those in the low-PBD, high-WSD, and non-smoker groups. In conclusion, genetic variants related to immunity and thyroid hormone secretion were linked to hypothyroidism risk, and their PRS interacted with PBD and WSD intake and smoking status. These results contribute to a better understanding of hypothyroidism and its prevention strategies for precision medicine intervention.

Keywords: hypothyroidism; white blood cell counts; immunity; inflammation; diet pattern

1. Introduction

Thyroid hormone production is tightly controlled by the hypothalamus–pituitary–thyroid (HPT) axis. When triiodothyronine (T3) and thyroxine (T4) levels are sufficient, they exert negative feedback on the hypothalamus and pituitary, reducing the release of thyrotropin-releasing hormone (TRH) and thyroid-stimulating hormone (TSH), respectively, to normalize T3 and T4 levels [1]. The disruption of the HPT axis elevates the TSH concentrations above the reference level and free thyroxine concentration below the reference range, leading to hypothyroidism. With a prevalence of approximately 5% in the general population, primary hypothyroidism accounts for over 99% of hypothyroid cases [1]. Hypothyroidism is more common in women and tends to increase with age [2].

Primary hypothyroidism typically results from damage to the thyroid gland itself, often due to autoimmune conditions such as Hashimoto's thyroiditis, iodine deficiency, certain medications, and prior thyroid surgery or radiation in persons having sufficient iodine intake [3]. The thyroid hormone regulates energy metabolism, body temperature, heart rate, and overall metabolism. Inadequate thyroid hormone levels lead to fatigue, weight gain, sensitivity to cold, dry skin, and cognitive impairment [4,5].

While the etiology of hypothyroidism is multifaceted and often complex, genetic factors have emerged as significant contributors to disease susceptibility. Recent genetic studies have shed light on the role of polygenic variants, which result from multiple genetic alterations across various loci, in influencing an individual's risk of developing hypothyroidism, particularly congenital hypothyroidism [6]. Genetic mutations associated with hypothyroidism are related to thyroid dysgenesis-related genes, such as *thyroid-stimulating hormone receptor (TSHR)*, *Forkhead Box E1 (FOXE1)*, *NK2 Homeobox 1 (NKX2-1)*, *Paired* Box 8 *(PAX8)*, and *NK2 Homeobox 5 (NKX2-5)*, as well as thyroid dyshormonogenesis linked genes, such as *Solute Carrier Family 5 Member 5 (SLC5A5)*, *Thyroid Peroxidase (TPO)*, *Dual oxidase 2 (DUOX2)*, *Dual oxidase maturation factor 2 (DUOXA2)*, *Solute Carrier Family 6 Member 4 (SLC6A4)*, and *iodothyronine dehalogenase (DEHAL1)* [6,7]. These genes are linked to essential thyroid functions, including thyroid differentiation, iodide organification, thyroglobulin synthesis, iodide transport, and iodotyrosine deiodination [6,8]. Understanding these genetic predispositions can offer valuable insights into disease pathogenesis, enabling early detection and personalized treatment approaches.

In addition to genetics, environmental factors have a crucial impact on thyroid hormone production, secretion, and function, contributing to the development of hypothyroidism [9]. These environmental influences can directly affect the thyroid gland or interfere with the production and regulation of TSH and thyroxin [9]. The factors include iodine deficiency, exposure to toxins, radiation, chronic stress, psychological factors affecting the HPT axis, consumption of certain goitrogenic foods (e.g., cabbage, broccoli, and cauliflower), exposure to endocrine-disrupting chemicals found in plastics and pesticides, and triggers related to autoimmune conditions such as viral infections and specific pollutants [10]. Despite the significance of these environmental factors, few studies have explored the relationship between hypothyroidism and dietary patterns. A review paper suggested that the Mediterranean diet may improve thyroid function [11]. However, the interaction between genetic variants and dietary intake in hypothyroidism remains largely unexplored.

Understanding the complex interplay between genetic susceptibility to hypothyroidism and the influence of dietary choices is vital for advancing our knowledge of disease pathogenesis. This study aimed to investigate the association of polygenic variants with hypothyroidism risk and the interaction of polygenic risk scores (PRS) with dietary patterns and lifestyle factors influencing disease risk. The findings of this research could potentially lead to the development of targeted and individualized approaches to managing and mitigating the risk of hypothyroidism.

2. Methods

2.1. Participants

This present study recruited 58,701 volunteers aged between 40 and 79 years for a cohort that included multiple hospitals as part of the Korean Genome and Epidemiology Study (KoGES) during 2010–2014 [8]. The institutional review boards (IRB) of the Korea National Institute of Health and Hoseo University approved the KoGES (KBP-2015-055 and 1041231-150811-HR-034-01, respectively). All participants signed a written informed consent form.

2.2. Baseline Characteristics, Anthropometric, and Biochemical Parameters of the Participants

General characteristics, including age, gender, education, income, smoking status, alcohol consumption, and physical activity, were surveyed on the initial visit [12,13]. Smoking status was categorized into non-smokers, past smokers who had not smoked for the last six months, and current smokers who had smoked at least 20 cigarettes in their lifetime [14]. The amount of alcohol intake was calculated by multiplying the alcohol content consumed on each occasion by frequency. Coffee intake was also measured in the same manner [14]. Regular physical activity was assessed by asking if the subjects engaged in 150 min or more of physical activity per week, and the response was marked "Yes" or "No".

Height and weight were obtained at the initial interview as previously described, and body mass index (BMI) was calculated by body weight (kg) divided by squares of height (m^2). Obesity was defined as BMI $\geq$ 25 kg/m^2. Blood pressure was determined by a physician using a sphygmomanometer in a sitting position under resting conditions three times, and the average systolic blood pressure (SBP) and diastolic blood pressure (DBP) were used. Blood was collected in a fasting state, and lipid profiles, alanine aminotransferase (ALT), aspartate aminotransferase (AST), alkaline phosphatase (ALP), and creatinine in the serum were measured using a Hitachi 7600 Automatic Analyzer (Hitachi, Tokyo, Japan). Plasma glucose and blood glycosylated hemoglobin (HbA1c) levels and white blood cell (WBC) count were measured using an automatic analyzer (ZEUS 9.9; Takeda, Tokyo, Japan). Serum C-reactive protein (CRP) concentrations were measured using an enzyme-linked immunosorbent assay (ELISA) kit.

2.3. Definition of Hypothyroidism and Metabolic Syndrome

Those participants diagnosed with hypothyroidism by a physician were considered to have the disease. The subjects with the experience of hyperthyroidism, cancers, chronic kidney disease, and brain-related diseases such as dementia and Parkinson's disease (n = 2037) were excluded from the study. The participants with and without hypothyroidism were 870 and 55,794, respectively. Metabolic syndrome (MetS) was defined as having three or more of the following traits: abdominal obesity measured by waist circumference, hyperglycemia, hypertension, hypo–high density lipoprotein (HDL) cholesterolemia, and hypertriglyceridemia, including conditions for which the individual may be taking medication. This definition was per the 2005 revised National Cholesterol Education Program Adult Treatment Panel III (NCEP-ATP III) criteria for Asia [15,16].

2.4. Estimation of Usual Food Intake by a Semi-Quantitative Food Frequency Questionnaire (SQFFQ)

An SQFFQ was designed to assess the usual food consumption of Koreans, and the accuracy and reproducibility were validated [17]. The SQFFQ included 106 food items Koreans consume as a typical diet. During the last year, the food intake was scored as more than, equal to, or less than the standard portion size in grams that was visualized using photographs of each food. Food frequencies were divided into the following nine categories: never or seldom, once per month, two to three times monthly, once or twice weekly, three or four times weekly, five or six times weekly, daily, twice daily, and $\geq$3 times daily. The daily food intakes were estimated by multiplying the median of the daily consumption frequencies by the portion size for each food category. The nutrients in the food intake per the SQFFQ were calculated using the Can-Pro 2.0 nutrient intake assessment software developed by the Korean Nutrition Society (Seoul, Republic of Korea). The dietary inflammation index was calculated by multiplying the inflammatory scores of 38 food components and nutrients reported in the previous study. However, garlic, ginger, saffron, and turmeric were excluded from the dietary inflammation index formula since their intake was not recorded in the KoGES study. The sums of the scores of 34 items were divided by 100, as described previously [18].

2.5. Dietary Patterns by Principal Component Analysis

Dietary pattern analysis was conducted based on the consumption of 30 predefined food groups from 106 food items in the SQFFQ, as previously reported [19]. Dietary patterns were generated using the pre-categorized food groups' principal component analysis (PCA). The number of clusters was made based on eigenvalues > 1.5, and four dietary patterns were generated [20]. The orthogonal rotation procedure (varimax) yielded four uncorrelated dietary patterns. Foods with ≥0.40 factor-loading values were considered to have a predominant contribution to the specific pattern [21]. The four dietary patterns were the Korean balanced diet (KBD), plant-based diet (PBD), Western-style diet (WSD), and rice-based diet (RBD) (Supplemental Table S1).

2.6. Quality Control of Genotyping and GWAS for Hypothyroidism Risk

The genotypes of the 58,701 participants were provided by the Center for Genome Science, Korea National Institute of Health. The genotyping of genomic DNA extracted from whole blood was performed using the Affymetrix Genome-Wide Human single nucleotide polymorphism (SNP) Array 5.0 (Affymetrix, Santa Clara, CA, USA). Genotyping quality and accuracy were determined using the Mahalanobis Distance genotyping algorithm with Bayesian robust linear modeling (BRLM) [18]. The genotype inclusion criteria were as follows: ≥98% genotyping accuracies, <4% rate of missing genotype calls, ≤30% heterozygosity, no gender biases, and $p > 0.05$ Hardy–Weinberg equilibrium (HWE) [18].

The genetic variant selection for hypothyroidism is present in Figure 1. The genetic association of hypothyroidism was investigated using genome-wide association studies (GWAS) after adjusting for age, gender, residence area, income, BMI, energy and alcohol intake, physical exercise, and smoking using PLINK (http://pngu.mgh.harvard.edu/~purcell/plink accessed on 10 January 2023). The distribution and appropriateness of genetic variants were checked using Manhattan and Quantile–Quantile (Q–Q) plots. The lambda value of the Q–Q plot was close to 1, indicating the appropriateness of the GWAS results. Among the genetic variants of the GWAS, those with high linkage disequilibrium (LD) ($D' < 0.2$, $r^2 > 0.05$) were excluded using Haploview 4.2 in PLINK since they provided the same genetic information for hypothyroidism risk [22].

2.7. Genetic Variant–Genetic Variant Interaction by a Generalized Multifactor Dimensionality Reduction (GMDR) Method

GMDR is a nonparametric and genetic model designed to detect and characterize nonlinear interactions between discrete genetic attributes. We applied the GMDR method to find the interacting genetic variants associated with hypothyroidism risk. The criteria for selecting the optimal genetic model were a significant p-value ($p < 0.05$) for the sign test of trained balance accuracy (TRBA) and test balance accuracy (TEBA) and a cross-validation consistency (CVC) score of 9 or 10 out of 10 [22]. The PRS was calculated by summing the number of the risk alleles in the genetic variants of the selected optimal model.

2.8. Statistical Analysis

Statistical analysis was performed using SAS (version 9.3; SAS Institute, Cary, NC, USA). The PRS of the selected model were categorized into three groups (low, middle, and high). The frequency distributions of the categorical variables were examined by applying the Chi-square test. A one-way analysis of variance (ANOVA) was conducted for the continuous variables among the PRS groups after adjusting for age, sex, BMI, education, income, energy intake, alcohol intake, smoking, and physical activity.

The association of hypothyroidism risk with anthropometric, biochemical, and genetic parameters was evaluated using adjusted logistic regression analysis. According to covariates, the first model was analyzed with adjustments for the area of residence, gender, age, and BMI, and the second model included the covariates in model 1 plus energy intake, smoking and drinking status, total physical activity, medication for asthma, and energy intake. The odds ratios (ORs) and 95% confidence intervals (CIs) were assessed using adjusted logistic regression using the low-PRS as a reference.

The potential interaction between the PRS and lifestyle factors for hypothyroidism risk was conducted with a multivariate general linear model (GLM) analysis with the main effects of PRS and lifestyle, their interaction, and covariates. Lifestyles were categorized into two groups (low and high levels) with the proper cutoff. An adjusted logistic regression analysis was performed in two groups based on the cutoff values assigned to each lifestyle parameter. The proper cutoff values for the two groups were assigned for each variable. This classification was based on the assumption that the low level of each parameter had a higher likelihood of interacting with the PRS. The specific cutoff values for each parameter were used with recommended intake for nutrients and 33rd percentiles of some parameters, such as dietary inflammation index and dietary patterns, and the exact values were found in the table legend. Based on the classification criteria, participants were then categorized into the high and low groups of lifestyle parameters. A p-value of ≤ 0.05 was considered statistically significant.

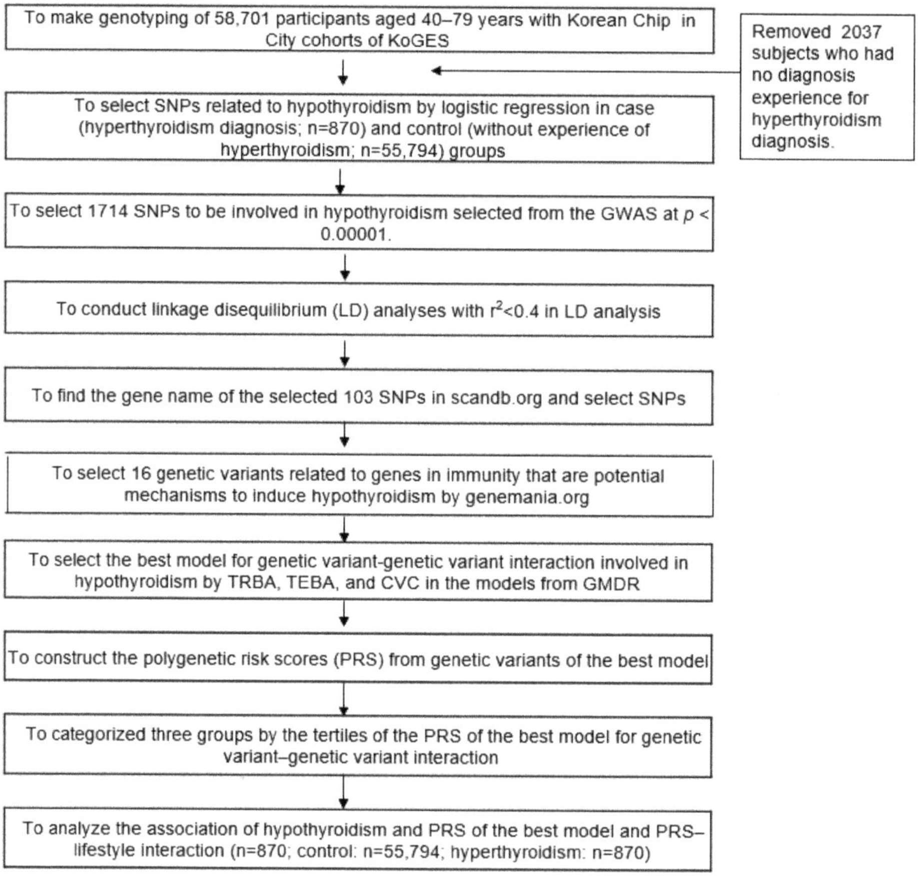

Figure 1. Flow chart to generate the polygenic risk score system influencing hypothyroidism risk.

3. Results

3.1. Demographic Characteristics and Nutrient Intake of Participants

The incidence of hypothyroidism in the study population was 1.48% (n = 870) (Table 1). Most participants with hypothyroidism were female; women had an 8.3-fold higher risk of hypothyroidism. Smoking status was not associated with hypothyroidism risk, but drinking was inversely associated with the risk of hypothyroidism (Table 1). The proportion of participants with hypothyroidism was higher in the exercise group than in the non-exercise group, but physical exercise was not significantly related to hypothyroidism risk. Intake of alcohol and coffee was lower in the participants with hypothyroidism than in healthy participants but only in women, and these intakes were inversely associated with hypothyroidism risk (Table 1). Energy intake did not differ between the healthy and the hypothyroid groups. Intake of carbohydrates, fats, proteins, and fiber did not differ in those with and without hypothyroidism. Iodine intake, one of the risk factors for hypothyroidism, did not vary between the healthy and hypothyroidism groups. Intake of antioxidant minerals, such as selenium, copper, and manganese, was much lower in the hypothyroidism group than the healthy group and was inversely associated with hypothyroidism (Table 1). However, the intake of antioxidant vitamin C did not differ between the two groups. Vitamin D intake and the dietary inflammation index did not differ between the two groups (Table 1).

Table 1. Demographic characteristics and nutrient intake of the participants according to hyperthyroidism.

	Men		Women		Adjusted ORs and 95% CI
	Normal (n = 19,970)	Hypothyroidism (n = 60)	Normal (n = 35,824)	Hypothyroidism (n = 810)	
Age (years)	54.8 ± 0.07 [a]	57.1 ± 1.23 [a]	49.6 ± 0.24 [b]	51.6 ± 1.71 [a***+++]	1.444 (1.243–1.678)
Gender (%)	35.8	6.9	64.2	93.1 ***	8.276 (6.294–10.88)
Hypothyroidism treatment (N, % treatment)	-	49 (81.7)	-	609 (75.2)	
Former+current Smoking (Number, %)	5604 (28.1)	14 (23.3)	702 (1.97)	13 (1.60)	0.859 (0.605–1.220)
Drinking (>20 g/day)	14,262 (71.6)	39 (65.0)	10,879 (30.5)	191 (23.6) ***	0.716 (0.603–0.851)
Coffee (>1 c/weeks)	13,510 (67.7)	41 (68.3)	21,292 (59.4)	444 (54.8) **	0.859 (0.74–0.990)
Physical activity (N, Yes%)	8822 (45.5)	33 (55.0)	13,426 (38.5)	354 (43.8) **	1.135 (0.986–1.306)
Energy (EER%) [1]	84.6 ± 0.05 [c]	84.3 ± 0.69 [c]	102 ± 0.03 [b]	103 ± 0.19 [a+++]	0.980 (0.853–1.127)
CHO (En%) [2]	60.9 ± 0.07 [b]	60.6 ± 1.09 [b]	72.5 ± 0.05 [a]	73.1 ± 0.29 [a+++]	1.152 (0.952-1.393)
Fat (En%) [3]	11.6 ± 0.05 [b]	11.4 ± 0.72 [b]	15.0 ± 0.03 [a]	15.0 ± 0.19 [a***]	0.516 (0.253–1.052)
Protein (En%) [4]	11.1 ± 0.02 [b]	11.3 ± 0.37 [b]	14.1 ± 0.02 [a]	14.1 ± 0.10 [a+++]	0.952 (0.779–1.164)
Fiber (g/day) [5]	5.59 ± 0.02 [b]	5.69 ± 0.34 [ab]	6.04 ± 0.07 [a]	6.08 ± 0.47 [ab++]	0.851 (0.703–1.031)
Iodine (ug/day) [6]	329 ± 2.8 [b]	317 ± 39.7 [b]	442 ± 1.89 [a]	450 ± 10.6 [a+++]	0.943 (0.809–1.100)
Selenium [7]	17.3 ± 0.19 [a]	11.1 ± 2.71 [ab]	13.0 ± 0.13 [b]	11.4 ± 0.72 [b**]	0.737 (0.616–0.882)
Cu [8]	0.93 ± 0.01 [a]	0.80 ± 0.10 [a]	0.87 ± 0.01 [a]	0.79 ± 0.03 [b*]	0.766 (0.642–0.914)
Mn [9]	2.27 ± 0.02 [a]	1.77 ± 0.21 [b]	2.08 ± 0.01 [a]	1.88 ± 0.06 [b**]	0.743 (0.633–0.872)
Zn [10]	4.74 ± 0.03	3.96 ± 0.40	4.72 ± 0.02	4.38 ± 0.11 **	0.810 (0.688–0.954)
Vitamin C [11]	91.4 ± 0.52 [b]	94.0 ± 7.45 [b]	113 ± 0.35 [a]	110 ± 1.99 [a+++]	1.053 (0.905–1.226)
Vitamin D [12]	5.21 ± 0.05 [b]	5.07 ± 0.66 [b]	7.07 ± 0.03 [a]	7.17 ± 0.18 [a+++]	1.050 (0.883–1.249)
Dietary inflammation index [13]	−18.5 ± 0.13 [b]	−19.1 ± 1.85 [ab]	−20.7 ± 0.09 [a]	−20.3+0.49 [a]	0.961 (0.812–1.137)
Sodium [14]	2.41 ± 0.01	2.39 ± 0.15	2.44 ± 0.01	2.35 ± +0.04	1.011 (0.871–1.174)
Seaweeds (g/day) [15]	1.66 ± 0.01 [b]	1.89 ± 0.27 [b]	2.45 ± 0.05 [a]	2.39 ± 0.37 [ab++]	0.868 (0.739–1.020)
Vegetables (g/day) [16]	89.1 ± 0.81 [b]	99.4 ± 11.5 [ab]	114 ± 0.54 [a]	115 ± 3.07 [a+]	0.973 (0.834–1.135)
Fruits (g/day) [17]	167 ± 1.9 [b]	165 ± 26.5 [b]	243 ± 1.3 [a]	240 ± 7.1 [a++]	1.080 (0.921–1.266)
Meats (g/day) [18]	86.7 ± 0.88 [a]	79.5 ± 12.5 [ab]	83.1 ± 0.60 [a]	75.0 ± 3.35 [b*]	0.803 (0.674–0.956) *
Traditional balanced diet [19]	14,661 (73.4)	47 (78.3)	22,151 (61.8)	458 (56.5) **	0.848 (0.731–0.984)
Prudent diet [19]	8943 (44.8)	28 (46.7)	26,455 (73.9)	636 (78.5) **	1.242 (1.045–1.476)
Western-style diet [19]	15,640 (78.3)	45 (75.0)	22,888 (63.9)	493 (60.9)	0.923 (0.794–1.073)
Rice-based diet [19]	12,790 (64.1)	37 (61.7)	23,848 (66.6)	513 (63.3)	0.877 (0.756–1.018)

Values represent adjusted means, standard errors, adjusted odds ratio (ORs), and 95% confidence intervals (CI) after adjusting for covariates of age, BMI, residence area, income, education, smoking and drinking status, and physical activity. In analyzing adjusted ORs, each parameter was divided into two groups with the cutoff, which was estimated energy requirement (EER) [1], 70 energy percent (En%) [2], 14 En% [3], 15 en% [4], 4 g/d [5], 461 ug/d [6], 15 ug/d [7], 1 ug/d [8], 2.1 ug/d [9], 5 ug/d [10], 100 mg/d [11], 10 ug/d [12]; 66th percentile [13], 2.0 g/d [14], 0.6 ug/d [15], 1 g/d [16], 2 g/d [16], 295 g/d [17], 76 g/d [18]; and 33th percentile [19]. Meats = meat + chicken + processed meat. Participants were divided into four groups according to hypothyroidism and gender. [a, b] Different superscript letters indicated significantly different at $p < 0.05$. * Significantly different by hypothyroidism at $p < 0.05$, ** at $p < 0.01$, and *** at $p < 0.001$. + Significantly different by genders at $p < 0.05$, ++ at $p < 0.01$, and +++ at $p < 0.001$.

Seaweed, vegetable, and fruit intake did not differ between the healthy and hypothyroidism groups (Table 1). Meat intake was lower in the hypothyroidism group than in the healthy group and was inversely linked to hypothyroidism risk. The proportion of participants with hypothyroidism was much lower in the high-KBD than in the low-KBD groups, and KBD was inversely associated with hypothyroidism risk (Table 1). However, PBD, WSD, and RMD were not associated with hypothyroidism risk (Table 1).

3.2. Anthropometric and Biochemical Measurements

BMI, waist and hip circumferences, plasma glucose, and blood HbA1c levels did not differ between the healthy and hypothyroidism groups and were not associated with its risk (Table 2). Serum total cholesterol and triglyceride concentrations were higher in the hypothyroidism group. The serum triglyceride concentrations were positively associated with hypothyroidism risk. The estimated glomerular filtration rate (eGFR, renal function index) and serum ALT and AST concentrations (liver function index) did not differ significantly between the two groups (Table 2). Interestingly, the serum ALP concentrations were much higher in the hypothyroidism group than in the healthy group, but only in women, and it was positively associated (2.11-fold) with hypothyroidism risk. There was a 3.15-fold increase in the risk of hypothyroidism in participants with <4.0 × 10^9 /L WBC count (Table 2). However, this association was not seen in those with serum CRP concentrations. Cancer incidence was higher in the hypothyroidism group than in the healthy group, and hypothyroidism risk was 1.93 times higher in those with cancer (Table 2). The cancers, primarily thyroid cancer, raised the risk of hypothyroidism 3.16 times. The incidence of osteoporosis and arthritis was higher in the hypothyroidism group than in the healthy group and was positively associated with hypothyroidism risk (Table 2).

Table 2. Adjusted means of the metabolic parameters according to gender and hypothyroidism.

	Men		Women		Adjusted ORs (95% CI)
	Normal (n = 19,970)	**Hypothyroidism (n = 60)**	**Normal (n = 35,824)**	**Hypothyroidism (n = 810)**	
BMI (kg/m^2) [1]	24.5 ± 0.02 ^{a}	24.3 ± 0.45 ^{a}	23.7 ± 0.09 ^{b}	23.5 ± 0.62 $^{a+\#}$	0.896 (0.765–1.048)
Waist circumference (cm) [2]	85.7 ± 0.04 ^{a}	86.2 ± 0.74 ^{a}	80.5 ± 0.15 ^{b}	79.8 ± 1.04 $^{b+++}$	0.850 (0.680–1.063)
Hip circumference (cm) [3] 98	95.7 ± 0.04 ^{a}	97.0 ± 0.67 ^{a}	94.4 ± 0.13 ^{b}	93.6 ± 0.93 $^{ab+++\#}$	1.007 (0.818–1.238)
Fasting plasma glucose (mg/dl) [4] 126	100 ± 0.2 ^{a}	99.0 ± 3.8 ^{ab}	97.1 ± 0.78 ^{b}	93.7 ± 5.33 ^{ab}	0.981 (0.740–1.300)
HbA1c (%) [5] 6.5	5.81 ± 0.01	5.75 ± 0.16	5.80 ± 0.04	5.61 ± 0.21	0.949 (0.667–1.348)
Total-C (mg/dl) [6] 230	193 ± 0.3 ^{b}	195 ± 5.8 ^{b}	201 ± 1.2 ^{a}	213 ± 8.1 $^{ab+}$	1.102 (0.939–1.292)
LDL-C (mg/dl) [7] 160	113 ± 0.3 ^{b}	112 ± 4.5 ^{b}	122 ± 0.2 ^{a}	118 ± 1.2 $^{ab++}$	1.142 (0.956–1.365)
HDL-C (mg/dl) [8]	49.1 ± 0.1 ^{b}	50.9 ± 1.9 ^{b}	56.9 ± 0.4 ^{a}	57.8 ± 0.5 $^{ab+++}$	0.967 (0.831–1.126)
TG (mg/dl) [9]	155 ± 0.9 ^{a}	159 ± 16.9 ^{ab}	125 ± 3.4 ^{b}	112 ± 23.6 $^{b++}$	1.274 (1.089–1.490) **
SBP (mmHg) [10]	125 ± 0.1 ^{a}	123 ± 2.2 ^{ab}	120 ± 0.4 ^{b}	120 ± 3.1 $^{ab+}$	0.878 (0.748–1.030)
DBP (mmHg) [11]	78.3 ± 0.1 ^{a}	77.0 ± 1.5 ^{b}	74.1 ± 0.3 ^{b}	71.9 ± 2.1 $^{b+++}$	0.751 (0.530–1.002)
MetS (N, Yes%)	16,430 (82.3)	47 (78.3)	31,433 (87.7)	713 (88.0)	1.078 (0.861–1.350)
eGFR [12] (mL/min/1.73m^2)	83.9 ± 0.14 ^{b}	86.6 ± 2.05 ab	87.3 ± 0.10 ^{a}	87.9 ± 0.55 $^{a+}$	0.862 0.693 1.074)
Serum ALT (U/L) [12]	26.7 ± 0.16 ^{a}	24.4 ± 3.0 ^{ab}	21.8 ± 0.61 ^{b}	22.4 ± 4.16 ^{ab}	0.856 (0.651–1.127)
Serum AST (U/L) [13]	25.7 ± 0.12 ^{a}	23.9 ± 2.23 ^{ab}	23.6 ± 0.46 ^{b}	25.2 ± 3.12 ^{ab}	1.090 (0.761–1.562)
Serum ALP (U/L) [14]	186 ± 1.2 ^{a}	172 ± 18.3 ^{ab}	167 ± 4.2 ^{b}	189 ± 25.3 ^{ab***}	2.110 (1.820–2.445)
WBC (×10^9 /L) [15] 4.0	5.80 ± 0.02 ^{a}	5.60 ± 0.20 ^{a}	5.65 ± 0.01 ^{ab}	5.55 ± 0.05 $^{b+}$	3.151 (2.601–3.817) ***
Serum CRP (mg/dL) [16] 0.5	0.17 ± 0.005	0.08 ± 0.10	0.13 ± 0.02	0.16 ± 0.15	1.355 0.959 1.914
Thyroid cancer (N, %)	35 (0.18)	0 (0)	336 (0.94)	24 (2.96) ***	3.160 2.074 4.816
Cancer incidence (N, Yes%)	536 (2.68)	1 (1.67)	1461 (4.08)	64 (7.90) ***	1.922 (1.480–2.495)
Asthma (N, Yes%)	279 (1.40)	0(0)	643 (1.80)	20 (2.47)	1.346 (0.857–2.112)
Osteoporosis (N, Yes%)	131 (0.66)	0(0)	2661 (7.43)	92 (11.4) ***	1.381 (1.089–1.752)
Arthritis (N, %)	787 (3.94)	3 (5.00)	3922 (11.0)	109 (13.5) *	1.241 (1.005–1.534)

Values represented adjusted means, standard error-adjusted odds ratios (ORs), and 95% confidence intervals (CI) after adjusting for covariates of age, BMI, residence area, income, education, smoking and drinking status, and physical activity. In analyzing adjusted ORs, each parameter was divided into two groups with the cutoff, which was 25 kg/m^2 [1], 90 cm for men and 85 cm for women [2]; 100 cm [3], 126 mg/dL [4], 230 mg/dL [5], 160 mg/dL [6], 45 for men and 50 mg/dL for women [8]; 150 mg/dL [9], 130 mmHg [10], 90 mmHg [11], 70 mL/min/1.73m^2 [12], 41 for men and 33 U/L for women [12]; 40 for men and 32 U/L for women [13]; and 129 for men and 104 U/L for women [14], 4.0 × 10^9 /L [15], 0.5 mg/dL [16]. ^{a,b} Different superscript letters indicated significantly different at $p < 0.05$. * Significantly different by hyperthyroidism at $p < 0.05$. ** at $p < 0.01$, and *** at $p < 0.001$. $^{+}$ Significantly different by gender at $p < 0.05$, $^{++}$ at $p < 0.01$, and $^{+++}$ at $p < 0.001$. $^{\#}$ Significant interaction between sex and hypothyroidism.

3.3. Genetic Variants Associated with Hypothyroidism Risk by GWAS and SNP–SNP Interactions by GMDR

The statistical significance of the genetic variants associated with hypothyroidism is shown in a Manhattan plot (Supplementary Figure S1A). Since the number of genetic variants with $p < 5 \times 10^{-8}$ was not sufficient to find proper genetic variants for hypothyroid risk, a liberal significance level ($p < 5 \times 10^{-5}$) was applied. Lambda, a genome inflation factor of genetic variants for hypothyroidism risk, was at 1.064, indicating no inflation of the genetic variants (Supplementary Figure S1B).

Among the genetic variants associated with hypothyroidism risk, ten genetic variants were selected with GMDR. The genetic characteristics of selected genetic variants ($p < 5 \times 10^{-5}$) are presented in Table 3. Three SNPs were positively associated (OR > 1), and seven SNPs were inversely associated with hyperthyroidism risk (0 < OR < 1). Two SNPs located in chromosome 6, including *Major Histocompatibility Complex, Class II, DQ Alpha 1 (HLA-DQA1)*, and *Chromosome 6 Open Reading Frame 15 (C6orf15)* and one SNP, *HORMA Domain Containing 2 (HORMAD2)*, in chromosome 22 were associated with immunity. One SNP in chromosome 8, *tumor necrosis factor receptor superfamily, member 11b (TNFRSF11B)*, was linked to inflammation (Table 3). Two SNPs, *Dual oxidase-1 (DUOX1)* and *Dual oxidase-2 (DUOX2)*, were associated with modulating oxidative stress in the thyroid, and *TSHR* was related to thyroid secretion. The minor allele frequency (MAF) of the ten SNPs was between 0.026 and 0.326. The p-value of the HWE was >0.05.

Table 3. The characteristics of the ten genetic variants related to an inflammation index used for the generalized multifactor dimensionality reduction analysis.

Chr [1]	SNP [2]	Position	Mi [3]	Ma [4]	OR [5]	*p* Value Adjusted [6]	MAF [7]	HWE [8]	Gene	Functional Consequence
1	rs144611984	108270345	A	C	1.91(1.48–2.46)	5.05×10^{-7}	0.0211	0.4238	*VAV3*	Intron
6	rs7990	32608077	A	C	1.37(1.23–1.54)	6.04×10^{-8}	0.19	0.0552	*HLA-DQA1*	Missense
6	rs28746784	32635140	T	C	1.48(1.28–1.71)	1.22×10^{-7}	0.0937	0.6098	*HLA-DQB1*	Nmd transcript
6	rs1800610	31543827	A	G	1.35(1.21–1.52)	2.13×10^{-7}	0.1921	0.6802	*TNF*	Intron
8	rs11573856	119954995	T	C	0.78(0.68–0.9)	4.74×10^{-5}	0.1826	0.1248	*TNFRSF11B*	Intron
11	rs11246015	224585	T	C	0.73(0.63–0.85)	5.52×10^{-6}	0.1482	0.2201	*SIRT3*	intron
12	rs7977554	112882859	A	G	1.55(1.31–1.83)	3.12×10^{-7}	0.0647	0.0878	*PTPN11*	Nmd transcript
14	rs75664963	81492195	T	A	0.77(0.68–0.86)	7.46×10^{-6}	0.2704	0.4389	*TSHR*	Intron
15	rs7171366	45386656	G	T	1.45(1.21–1.74)	4.83×10^{-6}	0.0567	0.1427	*DUOX2*	Intron
15	rs117742123	45429332	T	G	1.49(1.27–1.74)	5.82×10^{-7}	0.0786	1	*DUOX1*	Nmd transcript

[1] Chromosome; [2] Single nucleotide polymorphism; [3] Minor allele; [4] Major allele; [5] Odds ratio and lower and upper ends of 95% confidence interval; [6] *p*-value for OR after adjusting for age, gender, residence area, survey year, body mass index, daily energy intake, education, and income; [7] Minor allele frequency; [8] *p* value for Hardy–Weinberg equilibrium.

After conducting the GMDR, the optimal SNP–SNP interaction models contained three and seven genetic variants. The three-SNP model included *DUOX1*_rs1648314, *TSHR*_rs75664963, and *HLA-DQA1*_rs17426593. The seven-SNP model contained the SNPs in the three-SNP model plus *Sirtuin 3(SIRT3)* rs11246015, *Vav Guanine Nucleotide Exchange Factor 3 (VAV3)*_rs4915077, *TNFRSF11B*_rs1157, and *C6orf15*_rs2233955 (Supplementary Table S1). These three- and seven-SNP models had a *p*-value < 0.05 of the sign test of trained balanced accuracy (TRBA) and test balance accuracy (TEBA) after adjusting for age, gender, seaweed intake, and BMI for covariate set 1 and energy intake, physical activity, alcohol intake, and smoking status plus covariate set 1 for covariate set 2. The cross-validation consistency (CVC) of both models was 10/10 (Supplementary Table S1). These results indicated that the three- and seven-SNP models with SNP–SNP interactions contribute to the risk of hyperthyroidism.

3.4. Association between Polygenic Risk Scores (PRS) and Hypothyroidism Risk

The PRS was calculated from the three- or seven-SNP models by the GMDR models, and it was categorized into three groups: low-, medium-, and high-PRS. A high-PRS exhibited a 2.11 (1.63–2.74)- and 2.32 (1.48–3.33)-times higher hypothyroidism risk, respectively, compared with a low-PRS in the three-SNP and seven-SNP models after adjusting for the covariate set 2 (Figure 2). The adjusted ORs and 95% CI in the three-SNP and seven-SNP models were similar after adjustment of covariate sets 1 and 2. The PRS of the three-SNP model was associated with WBC counts and asthma but not serum CRP concentration and MetS-related biochemical parameters (Supplementary Table S2).

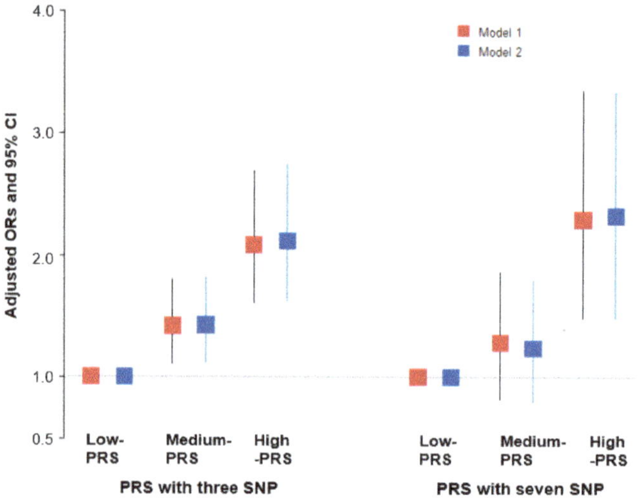

Figure 2. Adjusted odds ratio (ORs) and 95% confidence intervals (CIs) of the polygenic risk scores (PRS) of three- and seven-single nucleotide polymorphism (SNP) models generated for assessing SNP–SNP interactions associated with hypothyroidism risk. The best generalized multifactor dimensionality reduction analysis (GMDR) models with three-SNPs and seven-SNPs were calculated by summing the number of risk alleles of six and seven SNPs. The calculated PRS were divided into three categories (0–3, 4–5, and ≥6; 0–5, 6–8, and ≥9), the low-PRS, medium-PRS, and high-PRS groups, for the three-SNP and seven-SNP models, respectively. The adjusted OR was analyzed by logistic regression with covariates, including age, gender, residence areas, income, education, energy intake, smoking status, physical activity, alcohol intake, and the survey year. The reference group was the low-PRS in logistic regression. Red and blue boxes indicate the adjusted ORs for the three and seven SNPs, respectively, and the lines through red and blue boxes indicate 95% CIs.

3.5. Genetic Interactions of Lifestyle Factors with Hypothyroidism Risk

There was an interaction between the WBC count and the PRS from the three-SNP model to influence hypothyroidism risk. A high PRS increased the risk of hypothyroidism 4.89-fold in the low WBC count group and 1.71-fold in the high WBC count group (Figure 3A). The proportion of participants with hypothyroidism was much higher in the low WBC count than in the high WBC count group. The high-PRS group included more participants with hypothyroidism than the low-PRS group, regardless of WBC count. However, the serum CRP concentration did not influence hypothyroidism risk. The proportion of participants with hypothyroidism was much higher in the high CRP group than in the low CRP group (Figure 3B). The proportion of participants with hypothyroidism was higher in the high-PRS than the low-PRS group within the low CRP group. However, there was no significant difference among the PRS groups within the high CRP group (Table 4).

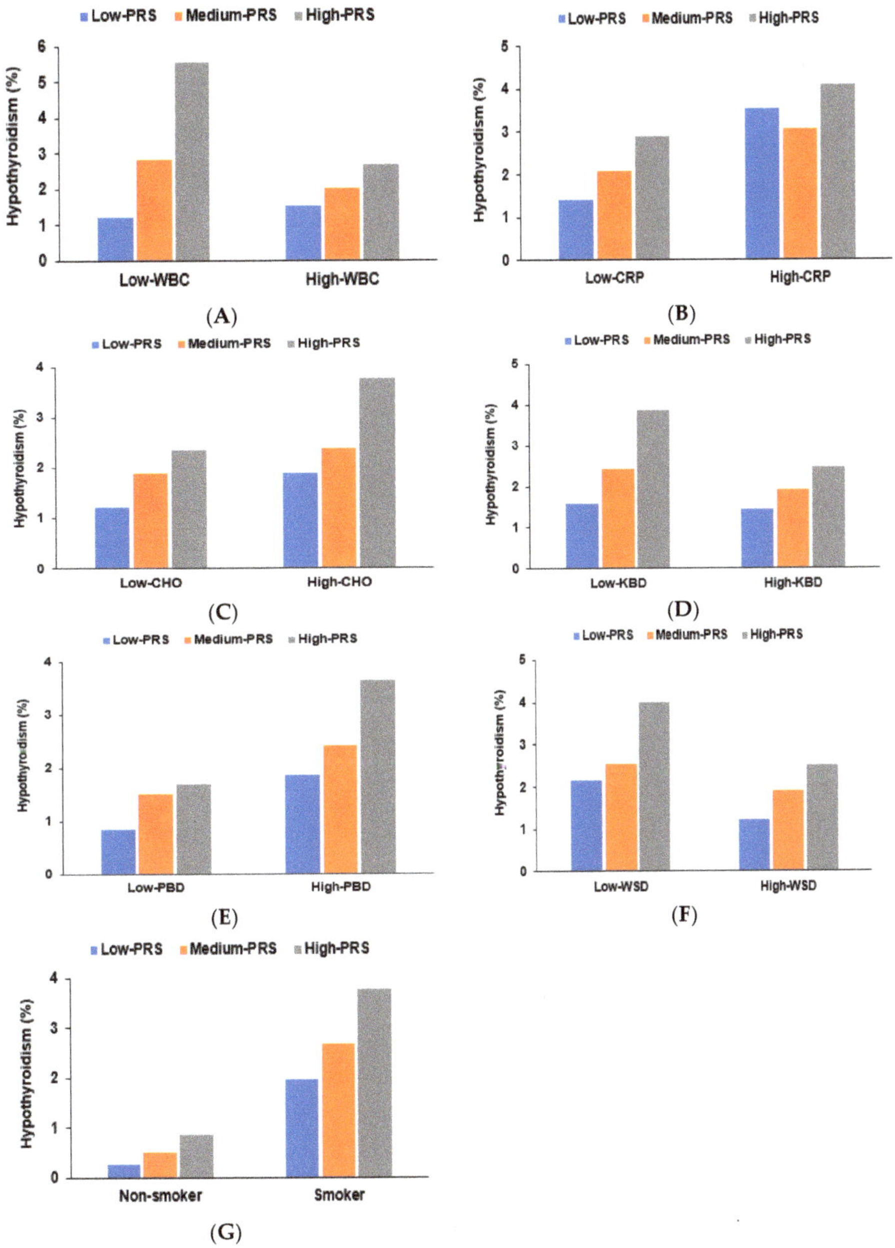

Figure 3. The proportion of individuals with hypothyroidism with the polygenic risk scores (PRS) of the three-single nucleotide polymorphism (SNP) model according to white blood cell (WBC) counts, diets, and smoking status. (**A**). WBC (cutoff: 4.0×10^9 /L); (**B**). Serum C-reactive protein (CRP) concentration (Cutoff: 0.5 mg/dL); (**C**). Carbohydrate intake (Cutoff: 70th energy percentile); (**D**). Korean balanced diet (KBD, Cutoff: 33rd percentile); (**E**). Plant-based diet (PBD, Cutoff: 33rd percentile); (**F**). Western-style diet (WSD, Cutoff: 33rd percentile); (**G**). Smoking status; Low-PRS (0–3), Medium-PRS (4–5), and High-PRS (≥6) in the six-SNP model.

Table 4. Adjusted odds ratios for the risk of serum CRP concentrations by the PRS with three SNPs[1] after covariate adjustments according to age, gender, metabolic syndrome, and nutrient intake.

	Low-PRS (N = 13,856)	Medium-PRS (N = 25,608)	High-PRS (N = 17,200)	Gene–Nutrient Interaction *p* Value
Low WBC [1]	1	2.299 (1.050–5.033)	4.887 (2.186–10.93)	<0.0001
High WBC		1.249 (0.942–1.655)	1.706 (1.257–2.315)	
Low CRP [2]	1	1.515 (1.174–1.954)	2.221 (1.693–2.915)	0.3053
High CRP	1	0.885 (0.306–2.560)	0.562 (0.209–1.510)	
Low EER [3]	1	1.248 (0.881–1.768)	1.893 (1.304–2.748)	0.2139
High EER	1	1.434 (0.945–2.178)	2.009 (1.281–3.149)	
Low CHO [4]	1	1.554 (1.088–2.220)	2.049 (1.396–3.009)	0.0563
High CHO	1	1.238 (0.860–1.782)	1.917 (1.297–2.834)	
Low protein [5]	1	1.315 (0.963–1.795)	1.852 (1.324–2.590)	0.2595
High protein	1	1.449 (0.938–2.237)	2.280 (1.438–3.614)	
Low fat [6]	1	1.285 (0.934–1.770)	1.762 (1.246–2.492)	0.1768
High fat 1	1	1.410 (0.865–2.297)	2.367 (1.415–3.958)	
Low KBD [7]	1	1.554 (1.034–2.335)	2.654 (1.731–4.067)	0.0608
High KBD	1	1.286 (0.912–1.812)	1.696 (1.167–2.464)	
Low PBD [7]	1	2.104 (1.184–3.739)	2.297 (1.235–4.274)	0.0140
High PBD	1	1.161 (0.865–1.558)	1.859 (1.357–2.547)	
Low WSD [7]	1	1.168 (0.797–1.712)	2.042 (1.360–3.065)	0.0295
High WSD	1	1.336 (0.957–1.865)	1.797 (1.254–2.577)	
Low RMD [7]	1	1.471 (0.969–2.234)	2.271 (1.461–3.531)	0.5219
High RMD	1	1.312 (0.944–1.824)	1.855 (1.299–2.648)	
Low PA [8]	1	1.285 (0.934–1.770)	1.762 (1.246–2.492)	0.7276
High PA	1	1.484 (1.020–2.159)	2.227 (1.495–3.319)	
Low coffee [9]	1	1.705 (1.088–2.671)	2.415 (1.498–3.894)	0.1439
High coffee	1	1.104 (0.790–1.541)	1.671 (1.165–2.397)	
Low alcohol [10]	1	1.419 (1.043–1.931)	1.962 (1.407–2.737)	0.2330
High alcohol	1	1.025 (0.597–1.760)	1.836 (1.041–3.237)	
Non-smokers	1	1.400 (1.089–1.800)	2.073 (1.584–2.711)	0.0379
Former and current Smokers	1	1.687 (0.504–5.645)	3.296 (0.952–11.41)	

Values represent odds ratios and 95% confidence intervals. Gene–gene interaction model with six SNPs included *CRP*_rs386636005, *GUSBP2*_rs1250561232, *OASL*_rs201853167, *APOC1*_rs56131196, *TLDC2*_rs59310406, and *HNF1A*_rs1169286. Low-GRS, medium-GRS, and high-GRS were divided into 0–4, 5–6, and >6 risk alleles for six SNP GMDR models, respectively. The cutoff points of dividing the values of each parameter into two groups were as follows: 4.0×10^9 /L [1], 0.5 mg/dL [2], estimated energy requirement (EER) [3]; 70 energy percent (En%) [4], 13 En% [5], 15 En% [6]; 66th percentile [7], 150 min/week of moderate-intensity physical activity [8], 3 cup/week [9], and 20 g/day [10]. Multivariate regression models include the corresponding main effects, interaction terms of gene and main effects (energy and nutrient intake), and potential confounders, such as age, gender, energy intake, residence area, metabolic syndrome, job, education, income, BMI, WBC, smoking, coffee, alcohol, and physical activity. The reference was the Low-PRS. KBD, Korean balanced diet; PBD, Plant-based diet; WSD, Western-style diet; RMD, rice-main diet; PA, physical activity.

Energy and macronutrient intakes did not interact with the PRS and did not contribute to hypothyroidism risk. However, carbohydrate intake was close to significant in the interaction with PRS ($p = 0.056$). The proportion of participants with hypothyroidism was much higher in those with a high carbohydrate intake than in those with a low intake (Figure 3C). The proportion of participants with hypothyroidism increased with the PRS, which was much higher in the low-KBD than in the high-KBD groups (Figure 3D). The PBD and WSD interacted with PRS to influence hypothyroidism risk among the different dietary patterns. The proportion of participants with hypothyroidism increased with the PRS, which was much higher in the high-PBD than in the low-PBD group (Figure 3E).

KBD showed a trend similar to that of the WSD but WSD showed a pattern opposite to that of PBD. The proportion was higher in those with a low WSD intake than those with a high WSD intake (Figure 3F). Among the lifestyle factors, physical activity, coffee, and alcohol intake did not interact with the PRS to contribute to hypothyroidism risk. However, smoking interacted with PRS to influence hypothyroidism risk (Figure 3G). The proportion of participants with hypothyroidism increased with PRS in both smokers and non-smokers but was much higher in smokers than in non-smokers in each PRS group.

4. Discussion

This present study highlights the association between genetic variants and hypothyroidism risk in a hospital-based cohort of participants aged over 40 years. Among the 56,664 participants, 870 were diagnosed with hypothyroidism, while 55,794 did not have hypothyroidism. This study identified several genetic variants related to immunity and the thyroid hormone secretion associated with hypothyroidism. The PRS derived from these genetic variants were positively correlated with hypothyroidism risk. Furthermore, the PRS showed interactions with lifestyle factors, such as dietary patterns (PBD and WSD) and smoking status, influencing hypothyroidism risk. These findings provide valuable insight into the genetic basis and potential preventive strategies for hypothyroidism.

Hypothyroidism exhibits distinct gender-related patterns in its prevalence. It is generally more common in females than males [23], consistent with the findings of the present study. Estrogen is a significant factor influencing thyroid function, and its interaction with various factors makes the relationship complex. However, the exact mechanisms through which estrogen affects the thyroid are still under investigation. Estrogen has been linked to the production, release, and metabolism of thyroid hormones from the thyroid gland. It modulates the synthesis of TSH in the pituitary gland [24,25]. Estrogen can also impact the levels of thyroid-binding proteins in the bloodstream, which, in turn, affects the transport and availability of thyroid hormones for cellular uptake and metabolism [26]. Moreover, elevated estrogen levels have been positively associated with developing autoimmune thyroid disorders, such as Hashimoto's thyroiditis [25]. Hormonal fluctuations at different life stages, such as puberty, pregnancy, and menopause, can impact thyroid function differently among individuals, potentially contributing to an increased risk of thyroid disorders in females [24,27].

Hypothyroidism is primarily a condition of severe dietary iodine deficiency called goiter. In case there is a sufficient intake of seaweed and iodine-fortified salt, the primary cause of hypothyroidism is thyroiditis due to an autoimmune disorder (Hashimoto's thyroiditis) and thyroidectomy or radioactive iodine therapy [1]. In different studies, the mean iodine intake of Koreans varies between 200 and 550 ug/day, mainly from seaweed intake (66%), milk and dairy products (11%), and fish (9%) [28,29]. This present study showed an intake of about 325 ug/day for men and 445 ug/day for women, which was higher than the recommended intake (150 ug/day) but lower than 2400 ug/day, the upper limit of iodine in Korea DRI. Previous studies have reported that excess iodine intake is associated with hyperthyroidism or hypothyroidism [30,31]. In Koreans, the risk of non-immune-related hypothyroidism is significantly elevated with excess iodine intake ($\geq$750 ug/day) with a hazard ratio (HR) of 2.81 times (1.64–4.80) [32]. Individuals with predisposing thyroid diseases, such as autoimmune thyroiditis or thyroidectomy, are susceptible to iodine-induced hypothyroidism when they consume iodine-fortified salt and drinking water [30]. However, this present study did not show a significant difference in iodine intake between the hypothyroidism and healthy groups. Therefore, hypothyroidism is not related to iodine intake in Koreans whose intake is adequate.

Moderate alcohol consumption serves as a protective factor against various autoimmune diseases, including overt autoimmune hypothyroidism [33,34]. Aligning with prior research, this current study establishes an inverse connection between alcohol consumption and hypothyroidism. However, the precise mechanism behind this effect remains undisclosed. The impact of coffee consumption on hypothyroidism remains uncertain. Analysis of NHANES data from 2007 to 2012 reveals that the incidence of subclinical hypothyroidism is lower when coffee intake is <2 cups per day compared to ≥2 cups per day, and it tends to be lower in the <2 cups per day group than in those who abstain from coffee [35]. In contrast, the participants in our study exhibited significantly lower coffee intake compared to NHANES participants, and this diminished coffee consumption showed an inverse correlation with hypothyroidism. The relationship between coffee intake and hypothyroidism necessitates further investigation.

Thyroid hormones cause the breakdown of carbohydrates and lipids for energy production and regulate the synthesis and breakdown of glucose, cholesterol, and triglycerides. It is suggested that thyroid hormones have multiple effects on glucose and lipid metabolism and energy consumption and play an essential role in MetS development. These hormones are also involved in heart function, the central nervous system, bone growth and turnover, and menstrual cycle and fertility in women. It has been observed in earlier studies that due to decreased thermogenesis and metabolic rates, individuals with hypothyroidism were more obese than healthy adults [36,37]. However, this present study showed that the BMI and waist circumferences were not significantly different between individuals with and without hypothyroidism. Furthermore, type 2 diabetes mellitus reduces TSH levels, impairs the conversion of T4 to T3 in the peripheral tissues, and is positively associated with hypothyroidism [38]. However, no association was observed between serum glucose concentration and hypothyroidism in this present study. Hypothyroidism has been reported to be linked to lipid profiles, but this remains controversial [39,40]. This present study showed that serum total cholesterol, HDL, and LDL concentrations were not associated with hypothyroidism, but serum triglyceride concentration was positively associated with hypothyroidism. Therefore, the MetS status was not linked to hypothyroidism risk in this present study.

Thyroid hormones have multiple effects on glucose and lipid metabolism and energy consumption and play an essential role in the development of MetS. Diseases such as thyroid cancer, osteoporosis, and arthritis are positively associated with hypothyroidism risk. WBC count indicates the immune status, and a low WBC count raises the risk of hypothyroidism 3.15-fold, indicating that hypothyroidism might be linked to autoimmune diseases. This might account for the higher incidence of most cancers, particularly thyroid cancer, in individuals with hypothyroidism. Previous studies have supported the association of cancer incidence with hypothyroidism risk. However, these results are inconsistent [41]: Breast cancer and radiation therapy to the supraclavicular lymph nodes are linked to hypothyroidism risk [42]. Elevated TSH, increased reactive oxygen species production, and mutation/polymorphisms of genes have been involved in increased cancer risk or pro-tumoral cell behavior [43]. Therefore, there is a need for more clinical studies in subclinical or clinical hypothyroidism to explore the relationship between lifestyle modification and hypothyroidism prevention.

The relationship between nutrient and diet intake and hypothyroidism remains unclear. Selenium and iodine are essential for thyroid hormone production and function, as their intake protects the thyroid gland from free radicals during thyroid hormone synthesis [44]. An increase in oxidative stress results in an oxidant/antioxidant imbalance in individuals with hypothyroidism [45,46]. A deficiency in selenium, zinc, and magnesium intake can significantly impact hypothyroidism risk. In a meta-analysis of 32 observational studies, individuals with hypothyroidism were observed to have lower serum selenium and zinc levels than healthy adults. [47]. A 10-week zinc, vitamin A, and magnesium supplementation intervention increases serum-free T4 concentration and prevents the increase of serum CRP and malondialdehyde levels and body weight in adults with hypothyroidism [48]. Selenium and zinc supplementation has been found to be beneficial in specific populations with otherwise limited generalizability [49]. This present study found that the intakes of selenium, copper, manganese, and zinc, which function as cofactors of antioxidant enzymes, were inversely associated with hypothyroidism risk, suggesting that hypothyroidism is linked to oxidative stress. However, the DII, vitamin C, vitamin D, and fiber intakes were not linked to hypothyroidism risk. Meat intake was inversely associated with hypothyroidism risk, but seaweed, vegetable, and fruit intakes were not associated with hypothyroidism. Furthermore, KBD was inversely associated, and PBD was positively associated with hypothyroidism risk. Therefore, maintaining a well-balanced diet that includes a variety of nutrient-rich foods is essential for supporting overall health and thyroid function in individuals with hypothyroidism.

Autoimmune activation can induce both hypothyroidism and hyperthyroidism, which result from different mechanisms within the autoimmune process [50]. Over time, the autoimmune attack leads to destroying thyroid tissue, reducing the gland's ability to produce thyroid hormones (T3 and T4) and inducing hypothyroidism [51]. However, in Graves' disease, a common autoimmune cause of hyperthyroidism, the immune system produces thyroid-stimulating immunoglobulins (TSI) that mimic the action of the TSH [51,52]. TSI binds to the TSH receptors on thyroid cells, stimulating the thyroid gland to overproduce thyroid hormones [52]. Therefore, both hypothyroidism and hyperthyroidism are associated with autoimmune disorders, and they are linked to immune and inflammation-related genetic variants. This present study showed that *HLA-DQA1*_rs17426593 and *TNFRSF11B*_rs11573856 were associated with hypothyroidism, suggesting a link between hypothyroidism, autoimmune disorders, and inflammation.

Thyroid hormone secretion is regulated by TSH acting via the TSH receptor (TSHR), a G protein-coupled transmembrane receptor. TSHR mediates thyroid hormone synthesis in the thyroid gland and height growth [53]. The TSHR loss-of-function mutation contributes to blocking TSH action to increase TSH levels to induce hypothyroidism with thyroid hypoplasia [53]. However, Graves' disease stimulates the production of TSHR antibodies, leading to the development of hyperthyroidism [52]. TSHR mutations are also reported to be associated with Graves' disease [51]. According to the location of mutation sites on the *TSHR* gene, hypothyroidism or hyperthyroidism may be induced. However, no *TSHR* mutation has been found for hyperthyroidism in the city hospital-based cohort of KoGES [50]. In Chinese patients with congenital hypothyroidism, seven genetic variants of *TSHR*, such as mutations in Ile216, Ala275, Asn372, and Ser567 with loss-of-function, and genetic variants of *DUOX2* are also found [54,55]. Therefore, the *TSHR* mutation may be mainly linked to hypothyroidism risk in Asians.

Notably, no previous studies have reported interactions between genetic variants and lifestyle factors and the association with hypothyroidism, making this present study unique and groundbreaking in its approach. This present study showed the interaction of genetic impact by PRS with PBD and WSD, contributing to hypothyroidism risk. Interestingly, the proportion of individuals with hypothyroidism increased with PRS and was much higher in the high PBD than in the low PBD groups. The trend of hypothyroidism with PRS in the PBD group was opposite to those in the WSD and KBD groups. The proportion of individuals with hypothyroidism was much higher in the low KBD and low WSD groups than in the high KBD and high WSD groups, regardless of the PRS. Smoking status had an interaction with the PRS to influence hypothyroidism risk. Overall, this study's findings demonstrated the intricate relationship between genetic factors, dietary patterns, and smoking status in the context of hypothyroidism risk. Identifying these interactions adds significant value to our existing knowledge and highlights the importance of considering genetic and lifestyle factors in understanding and managing hypothyroidism. Further research is needed to explore the underlying mechanisms and clinical implications of these interactions for developing personalized preventive strategies and targeted interventions for individuals at risk of hypothyroidism.

In this present study, a novel finding was the relationship of the PRS for hypothyroidism with immunity and thyroid function and its interaction with dietary patterns. However, the limitations of this study may be summarized as follows: (1) The data were derived from a cross-sectional study. Therefore, causal relationships could not be established. (2) Serum TSH, T3, and T4 concentrations were not measured, and the participants were classified as having hypothyroidism based on a question regarding any previous diagnosis of hypothyroidism. Misclassification or measurement errors in these variables could introduce bias into the observed associations. (3) The usual food intake was calculated from the SQFFQ containing 106 foods designed for Korean meals, which was validated with 3-day food records for four seasons. However, the SQFFQ has some measurement bias in measuring food intake. 4) The genetic variants were not validated in an independent cohort since other cohorts available in Korea did not include hypothyroidism-related data. However, the number of participants was large enough to enhance the statistical power for conducting robust analyses and increasing the generalizability of the results.

In conclusion, this hospital-based cohort study provides valuable insights into the genetic basis and potential preventive strategies for hypothyroidism. This study identified genetic variants related to immunity and thyroid hormone secretion that are associated with hypothyroidism risk. The PRS derived from these genetic variants was positively correlated with hypothyroidism risk, highlighting the importance of genetic factors in disease susceptibility. Moreover, this study demonstrated that the PRS interacts with certain lifestyle factors, specifically PBD and WSD, as well as smoking status, influencing hypothyroidism risk. Notably, the interaction of PRS with the PBD and WSD leads to contrasting trends in hypothyroidism risk, depending on the dietary patterns followed by individuals. These findings emphasize the intricate interplay between genetic and lifestyle factors in determining hypothyroidism risk, thus adding novel insights to our understanding of the condition.

Supplementary Materials: The following supporting information can be downloaded at: https://www.mdpi.com/article/10.3390/nu15173850/s1, Figure S1: The distribution of genetic variants related to hypothyroidism; Manhattan plot of genetic variants related to hypothyroidism; Q–Q plot of genetic variants related to hypothyroidism. Table S1: The characteristics of the ten genetic variants of genes in the risk of inflammation used for the generalized multifactor dimensionality. Table S2. Adjusted odds ratios for the risk of hypothyroidism by polygenetic risk scores of the 3 SNPs model (PRS) for gene-gene interaction after covariate adjustments reduction analysis.

Author Contributions: D.S.K., data quality assurance and analysis; S.P., data management, data analysis, interpretation of results, and writing. All authors have read and agreed to the published version of the manuscript.

Funding: This study was supported by a grant from the National Research Foundation of Korea (NRF) funded by the Ministry of Science and ICT (RS-2023-00208567).

Institutional Review Board Statement: The institutional review board (IRB) of the Korea National Institute of Health approved the KoGES (KBP-2015-055), and the IRB of Hoseo University accepted the use of the KoGES data to conduct the present study (HR-034-01).

Informed Consent Statement: All participants signed a written informed consent form.

Data Availability Statement: The data were deposited in the Korean biobank (Osong, Republic of Korea).

Acknowledgments: The authors acknowledge the participants and researchers involved in KoGES.

Conflicts of Interest: The authors declare no conflict of interest.

References

1. Chiovato, L.; Magri, F.; Carlé, A. Hypothyroidism in Context: Where We've Been and Where We're Going. *Adv. Ther.* **2019**, *36*, 47–58. [CrossRef]
2. Leng, O.; Razvi, S. Hypothyroidism in the older population. *Thyroid. Res.* **2019**, *12*, 2. [CrossRef]
3. Chaker, L.; Bianco, A.C.; Jonklaas, J.; Peeters, R.P. Hypothyroidism. *Lancet* **2017**, *390*, 1550–1562. [CrossRef]
4. Sawicka-Gutaj, N.; Erampamoorthy, A.; Zybek-Kocik, A.; Kyriacou, A.; Zgorzalewicz-Stachowiak, M.; Czarnywojtek, A.; Ruchała, M. The Role of Thyroid Hormones on Skeletal Muscle Thermogenesis. *Metabolites* **2022**, *12*, 336. [CrossRef]
5. Su, X.; Peng, H.; Chen, X.; Wu, X.; Wang, B. Hyperlipidemia and hypothyroidism. *Clin. Chim. Acta* **2022**, *527*, 61–70. [CrossRef] [PubMed]
6. Kostopoulou, E.; Miliordos, K.; Spiliotis, B. Genetics of primary congenital hypothyroidism—A review. *Hormones* **2021**, *20*, 225–236. [CrossRef] [PubMed]

7. Kizys, M.M.L.; Louzada, R.A.; Mitne-Neto, M.; Jara, J.R.; Furuzawa, G.K.; de Carvalho, D.P.; Dias-da-Silva, M.R.; Nesi-França, S.; Dupuy, C.; Maciel, R.M.B. DUOX2 Mutations Are Associated with Congenital Hypothyroidism With Ectopic Thyroid Gland. *J. Clin. Endocrinol. Metab.* **2017**, *102*, 4060–4071. [CrossRef]
8. Stoupa, A.; Kariyawasam, D.; Polak, M.; Carré, A. Genetics of congenital hypothyroidism: Modern concepts. *Pediatr. Investig.* **2022**, *6*, 123–134. [CrossRef]
9. Babić Leko, M.; Gunjača, I.; Pleić, N.; Zemunik, T. Environmental Factors Affecting Thyroid-Stimulating Hormone and Thyroid Hormone Levels. *Int. J. Mol. Sci.* **2021**, *22*, 6521. [CrossRef] [PubMed]
10. Gonzalez, M.L.; Chernock, R.D.; Mansour, M. Environmental factors and anatomic pathology of the thyroid gland: Review of literature. *Diagn. Histopathol.* **2020**, *26*, 207–215. [CrossRef]
11. Bellastella, G.; Scappaticcio, L.; Caiazzo, F.; Tomasuolo, M.; Carotenuto, R.; Caputo, M.; Arena, S.; Caruso, P.; Maiorino, M.I.; Esposito, K. Mediterranean Diet and Thyroid: An Interesting Alliance. *Nutrients* **2022**, *14*, 4130. [CrossRef] [PubMed]
12. Daily, J.W.; Park, S. Interaction of BDNF rs6265 variants and energy and protein intake in the risk for glucose intolerance and type 2 diabetes in middle-aged adults. *Nutrition* **2017**, *33*, 187. [CrossRef] [PubMed]
13. Park, S.; Daily, J.W.; Zhang, X.; Jin, H.S.; Lee, H.J.; Lee, Y.H. Interactions with the MC4R rs17782313 variant, mental stress and energy intake and the risk of obesity in Genome Epidemiology Study. *Nutr. Metab.* **2016**, *13*, 38. [CrossRef] [PubMed]
14. Park, S.; Ahn, J.; Lee, B.K. Self-rated Subjective Health Status Is Strongly Associated with Sociodemographic Factors, Lifestyle, Nutrient Intakes, and Biochemical Indices, but Not Smoking Status: KNHANES 2007–2012. *J. Korean Med. Sci.* **2015**, *30*, 1279–1287. [CrossRef]
15. Park, S.; Kim, D.S.; Kang, S. Carrying minor allele of FADS1 and haplotype of FADS1 and FADS2 increased the risk of metabolic syndrome and moderate but not low fat diets lowered the risk in two Korean cohorts. *Eur. J. Nutr.* **2019**, *58*, 831–842. [CrossRef]
16. Park, S.; Ahn, J.; Lee, B.K. Very-low-fat diets may be associated with increased risk of metabolic syndrome in the adult population. *Clin. Nutr.* **2016**, *35*, 1159–1167. [CrossRef]
17. Park, S.; Zhang, X.; Lee, N.R.; Jin, H.S. TRPV1 Gene Polymorphisms Are Associated with Type 2 Diabetes by Their Interaction with Fat Consumption in the Korean Genome Epidemiology Study. *J. Nutr. Nutr.* **2016**, *9*, 47–61. [CrossRef]
18. Hur, H.J.; Yang, H.J.; Kim, M.J.; Lee, K.H.; Kim, M.S.; Park, S. Association of Polygenic Variants with Type 2 Diabetes Risk and Their Interaction with Lifestyles in Asians. *Nutrients* **2022**, *14*, 3222. [CrossRef]
19. Ahn, Y.; Kwon, E.; Shim, J.E.; Park, M.K.; Joo, Y.; Kimm, K.; Park, C.; Kim, D.H. Validation and reproducibility of food frequency questionnaire for Korean genome epidemiologic study. *Eur. J. Clin. Nutr.* **2007**, *61*, 1435–1441. [CrossRef]
20. Steel, R.G.; Torrie, J.H.; Dickey, D.A. *Principles and Procedures of Statistics: A Biological Approach*; McGraw-Hill: New York, NY, USA, 1997.
21. Kim, J.O.; Mueller, C.W. *Factor Analysis. Statistical Methods and Practical Issues*; Sage publications: Thousand Oaks, CA, USA, 1978.
22. Hur, H.J.; Wu, X.; Yang, H.J.; Kim, M.J.; Lee, K.-H.; Hong, M.; Park, S.; Kim, M.-S. Beneficial Effects of a Low-Glycemic Diet on Serum Metabolites and Gut Microbiota in Obese Women with Prevotella and Bacteriodes Enterotypes: A Randomized Clinical Trial. *Front. Nutr.* **2022**, *9*, 861880. [CrossRef]
23. Davis, M.G.; Phillippi, J.C. Hypothyroidism: Diagnosis and Evidence-Based Treatment. *J. Midwifery Womens Health* **2022**, *67*, 394–397. [CrossRef]
24. Hapon, M.B.; Simoncini, M.; Via, G.; Jahn, G.A. Effect of hypothyroidism on hormone profiles in virgin, pregnant and lactating rats, and on lactation. *Reproduction* **2003**, *126*, 371–382. [CrossRef] [PubMed]
25. Redmond, G.P. Hypothyroidism and women's health. *Int. J. Fertil. Womens Med.* **2002**, *47*, 123–127. [PubMed]
26. Kaminski, J.; Junior, C.M.; Pavesi, H.; Drobrzenski, B.; Amaral, G.M.D. Effects of oral versus transdermal estradiol plus micronized progesterone on thyroid hormones, hepatic proteins, lipids, and quality of life in menopausal women with hypothyroidism: A clinical trial. *Menopause* **2021**, *28*, 1044–1052. [CrossRef]
27. Frank-Raue, K.; Raue, F. Thyroid Dysfunction in Periand Postmenopausal Women-Cumulative Risks. *Dtsch. Arztebl. Int.* **2023**, *120*, 311–316.
28. Kim, S.; Kwon, Y.S.; Kim, J.Y.; Hong, K.H.; Park, Y.K. Association between Iodine Nutrition Status and Thyroid Disease-Related Hormone in Korean Adults: Korean National Health and Nutrition Examination Survey VI (2013–2015). *Nutrients* **2019**, *11*, 2757. [CrossRef] [PubMed]
29. Lee, J.; Yeoh, Y.; Seo, M.J.; Lee, G.H.; Kim, C.-I. Estimation of Dietary Iodine Intake of Koreans through a Total Diet Study (TDS). *Korean J. Community Nutr.* **2021**, *26*, 48–55. [CrossRef]
30. Katagiri, R.; Yuan, X.; Kobayashi, S.; Sasaki, S. Effect of excess iodine intake on thyroid diseases in different populations: A systematic review and meta-analyses including observational studies. *PLoS ONE* **2017**, *12*, e0173722. [CrossRef]
31. Wu, K.; Zhou, Y.; Ke, S.; Huang, J.; Gao, X.; Li, B.; Lin, X.; Liu, X.; Liu, X.; Ma, L.; et al. Lifestyle is associated with thyroid function in subclinical hypothyroidism: A cross-sectional study. *BMC Endocr. Disord.* **2021**, *21*, 112. [CrossRef]
32. Kim, H.I.; Oh, H.K.; Park, S.Y.; Jang, H.W.; Shin, M.H.; Han, J.M.; Bae, J.C.; Kim, S.W.; Kim, T.H.; Chung, J.H. Non-immune-related hypothyroidism and its relationship with excess iodine. *Eur. J. Nutr.* **2019**, *58*, 2851–2858. [CrossRef]
33. Carlé, A.; Pedersen, I.B.; Knudsen, N.; Perrild, H.; Ovesen, L.; Rasmussen, L.B.; Jørgensen, T.; Laurberg, P. Moderate alcohol consumption may protect against overt autoimmune hypothyroidism: A population-based case-control study. *Eur. J. Endocrinol.* **2012**, *167*, 483–490. [CrossRef]

34. Wiersinga, W.M. Clinical Relevance of Environmental Factors in the Pathogenesis of Autoimmune Thyroid Disease. *Endocrinol. Metab.* **2016**, *31*, 213–222. [CrossRef]
35. Zhao, G.; Wang, Z.; Ji, J.; Cui, R. Effect of coffee consumption on thyroid function: NHANES 2007–2012 and Mendelian randomization. *Front. Endocrinol.* **2023**, *14*, 1188547. [CrossRef] [PubMed]
36. Yan, Y.; Xu, M.; Wu, M.; Wang, X.; Li, F.; Zhang, J.; You, L.; Pan, X.; Feng, W.; Wu, J.; et al. Obesity is associated with subclinical hypothyroidism in the presence of thyroid autoantibodies: A cross-sectional study. *BMC Endocr. Disord.* **2022**, *22*, 94. [CrossRef] [PubMed]
37. Song, R.-H.; Wang, B.; Yao, Q.-M.; Li, Q.; Jia, X.; Zhang, J.-A. The Impact of Obesity on Thyroid Autoimmunity and Dysfunction: A Systematic Review and Meta-Analysis. *Front. Immunol.* **2019**, *10*, 2349. [CrossRef] [PubMed]
38. Kalra, S.; Aggarwal, S.; Khandelwal, D. Thyroid Dysfunction and Type 2 Diabetes Mellitus: Screening Strategies and Implications for Management. *Diabetes Ther.* **2019**, *10*, 2035–2044. [CrossRef] [PubMed]
39. Liu, H.; Peng, D. Update on dyslipidemia in hypothyroidism: The mechanism of dyslipidemia in hypothyroidism. *J. Endocr. Connect.* **2022**, *11*, e210002.
40. Luo, Y.; Wu, F.; Huang, Z.; Gong, Y.; Zheng, Y. Assessment of the relationship between subclinical hypothyroidism and blood lipid profile: Reliable or not? *Lipids Health Dis.* **2022**, *21*, 137. [CrossRef] [PubMed]
41. Gómez-Izquierdo, J.; Filion, K.B.; Boivin, J.F.; Azoulay, L.; Pollak, M.; Yu, O.H.Y. Subclinical hypothyroidism and the risk of cancer incidence and cancer mortality: A systematic review. *BMC Endocr. Disord.* **2020**, *20*, 83.
42. Solmunde, E.; Falstie-Jensen, A.M.; Lorenzen, E.L.; Ewertz, M.; Reinertsen, K.V.; Dekkers, O.M.; Cronin-Fenton, D.P. Breast cancer, breast cancer-directed radiation therapy and risk of hypothyroidism: A systematic review and meta-analysis. *Breast* **2023**, *68*, 216–224. [CrossRef] [PubMed]
43. Penna, G.; Rubio, I.G.S.; Brust, E.S.; Cazarin, J.; Hecht, F.; Alkmim, N.R.; Rajão, K.M.A.B.; Ramos, H.E. Congenital hypothyroidism and thyroid cancer. *J. Endocr. Relat. Cancer* **2021**, *28*, R217–R230. [CrossRef] [PubMed]
44. Guastamacchia, E.; Giagulli, V.A.; Licchelli, B.; Triggiani, V. Selenium and Iodine in Autoimmune Thyroiditis. *Endocr. Metab. Immune Disord. Drug Targets* **2015**, *15*, 288–292. [CrossRef] [PubMed]
45. Ates, I.; Yilmaz, F.M.; Altay, M.; Yilmaz, N.; Berker, D.; Güler, S. The relationship between oxidative stress and autoimmunity in Hashimoto's thyroiditis. *Eur. J. Endocrinol.* **2015**, *173*, 791–799. [CrossRef]
46. Ruggeri, R.M.; CampennÌ, A.; Giuffrida, G.; Casciaro, M.; Barbalace, M.C.; Hrelia, S.; Trimarchi, F.; CannavÒ, S.; Gangemi, S. Oxidative stress as a key feature of autoimmune thyroiditis: An update. *Minerva Endocrinol.* **2020**, *45*, 326–344. [CrossRef]
47. Talebi, S.; Ghaedi, E.; Sadeghi, E.; Mohammadi, H.; Hadi, A.; Clark, C.C.T.; Askari, G. Trace Element Status and Hypothyroidism: A Systematic Review and Meta-analysis. *Biol. Trace Elem. Res.* **2020**, *197*, 1–14. [CrossRef]
48. Rabbani, E.; Golgiri, F.; Janani, L.; Moradi, N.; Fallah, S.; Abiri, B.; Vafa, M. Randomized Study of the Effects of Zinc, Vitamin A, and Magnesium Co-supplementation on Thyroid Function, Oxidative Stress, and hs-CRP in Patients with Hypothyroidism. *Biol. Trace Elem. Res.* **2021**, *199*, 4074–4083. [CrossRef] [PubMed]
49. Larsen, D.; Singh, S.; Brito, M. Thyroid, Diet, and Alternative Approaches. *J. Clin. Endocrinol. Metab.* **2022**, *107*, 2973–2981. [PubMed]
50. Song, M.Y.; Park, S. Association of Polygenetic Risk Scores Related to Immunity and Inflammation with Hyperthyroidism Risk and Interactions between the Polygenetic Scores and Dietary Factors in a Large Cohort. *J. Thyroid. Res.* **2021**, *2021*, 7664641. [CrossRef] [PubMed]
51. Stefan, M.; Faustino, L.C. Genetics of Thyroid-Stimulating Hormone Receptor—Relevance for Autoimmune Thyroid Disease. *Front. Endocrinol.* **2017**, *8*, 57. [CrossRef] [PubMed]
52. Giménez-Barcons, M.; Colobran, R.; Gómez-Pau, A.; Marín-Sánchez, A.; Casteràs, A.; Obiols, G.; Abella, R.; Fernández-Doblas, J.; Tonacchera, M.; Lucas-Martín, A.; et al. Graves' Disease TSHR-Stimulating Antibodies (TSAbs) Induce the Activation of Immature Thymocytes: A Clue to the Riddle of TSAbs Generation? *J. Immunol.* **2015**, *194*, 4199–4206.
53. Schoenmakers, N.; Chatterjee, V.K. TSHR mutations and subclinical congenital hypothyroidism. *Nat. Rev. Endocrinol.* **2015**, *11*, 258–259. [CrossRef] [PubMed]
54. Fang, Y.; Sun, F.; Zhang, R.J.; Zhang, C.R.; Yan, C.Y.; Zhou, Z.; Zhang, Q.Y.; Li, L.; Ying, Y.X.; Zhao, S.X.; et al. Mutation screening of the TSHR gene in 220 Chinese patients with congenital hypothyroidism. *Clin. Chim. Acta* **2019**, *497*, 147–152. [CrossRef] [PubMed]
55. Fu, C.; Wang, J.; Luo, S.; Yang, Q.; Li, Q.; Zheng, H.; Hu, X.; Su, J.; Zhang, S.; Chen, R.; et al. Next-generation sequencing analysis of TSHR in 384 Chinese subclinical congenital hypothyroidism (CH) and CH patients. *Clin. Chim. Acta* **2016**, *462*, 127–132. [CrossRef] [PubMed]

nutrients

MDPI

Article

Association of the DNA Methylation of Obesity-Related Genes with the Dietary Nutrient Intake in Children

Priyadarshni Patel [1], Vaithinathan Selvaraju [1], Jeganathan Ramesh Babu [1,2] and Thangiah Geetha [1,2,*]

1 Department of Nutritional Sciences, Auburn University, Auburn, AL 36849, USA
2 Boshell Metabolic Diseases and Diabetes Program, Auburn University, Auburn, AL 36849, USA
* Correspondence: thangge@auburn.edu; Tel.: +1-334-844-7418; Fax: +1-334-844-3268

Abstract: The occurrence of obesity stems from both genetic and external influences. Despite thorough research and attempts to address it through various means such as dietary changes, physical activity, education, and medications, a lasting solution to this widespread problem remains elusive. Nutrients play a crucial role in various cellular processes, including the regulation of gene expression. One of the mechanisms by which nutrients can affect gene expression is through DNA methylation. This modification can alter the accessibility of DNA to transcription factors and other regulatory proteins, thereby influencing gene expression. Nutrients such as folate and vitamin B12 are involved in the one-carbon metabolism pathway, which provides the methyl groups necessary for DNA methylation. Studies have shown that the inadequate intake of these nutrients can lead to alterations in DNA methylation patterns. For this study, we aim to understand the differences in the association of the dietary intake between normal weight and overweight/obese children and between European American and African American children with the DNA methylation of the three genes *NRF1*, *FTO*, and *LEPR*. The research discovered a significant association between the nutritional intake of 6–10-years-old children, particularly the methyl donors present in their diet, and the methylation of the *NRF1*, *FTO*, and *LEPR* genes. Additionally, the study emphasizes the significance of considering health inequalities, particularly family income and maternal education, when investigating the epigenetic impact of methyl donors in diet and gene methylation.

Keywords: childhood obesity; DNA methylation; epigenetics; health disparities; folate intake; methyl donors; dietary intake

1. Introduction

Childhood overweight and obesity results from consistently consuming more energy than needed and is influenced by a combination of genetics, lifestyle choices, the obesogenic environment, and social factors [1]. While there is evidence of a genetic component involved in childhood obesity, the significant rise in obesity rates among children cannot be solely attributed to genetic changes [2,3]. This suggests that interactions between genes and the environment are likely driving the epidemic of childhood obesity. Obesity, which is defined as the abnormal accumulation of excess body fat, can lead to various health issues such as blood lipid disorders, high blood pressure, insulin resistance, type 2 diabetes, metabolic syndrome, cardiovascular disease, and liver fat accumulation [4]. The prevalence of obesity is rapidly increasing in modern society, with an estimated 58% of adults worldwide expected to meet the criteria for obesity by 2030 [5]. The literature shows that approximately 55% of children who are obese will continue to be obese during their teenage years [6]. Roughly 80% of obese teenagers remain obese in adulthood, and approximately 70% stay obese beyond the age of 30 years [6]. When we look closer, obesity disproportionately affects racial minority groups, especially Hispanic and Black groups; therefore, it is important to understand the causes and reduce the prevalence of childhood obesity [7].

As the rise in childhood obesity rates in recent decades cannot be fully attributed to genetics alone, factors such as nutrition and lifestyle in our surroundings may also play a role in this trend. These factors can influence gene expression without altering the DNA sequence, a phenomenon known as epigenetics [8]. Epigenetics explores how external influences, such as lifestyle choices, exercise, toxin exposure, and diet, can impact gene expression [9–11]. These modifications play a role in various conditions, including obesity [12–14], type 2 diabetes [15], metabolic syndrome [16], insulin resistance [17], and cancer [18]. Conrad Waddington introduced the concept of epigenetics in 1942, which involves chemical modifications that affect how the body interprets DNA [19]. The most extensively researched epigenetic changes include DNA methylation, histone modifications, and non-coding RNAs [8]. Importantly, various therapeutic approaches, such as a low-calorie diet [20,21], bariatric surgery [22,23], and physical activity [24,25], can reverse these epigenetic markers that occur during obesity management. Nutrients can also serve as a source of epigenetic modifications and reverse specific disease-associated epigenetic markers [26,27]. As a result, nutritional epigenetics has emerged as a novel mechanism that explains how nutrition interacts with genes, providing evidence of its role in metabolic diseases.

Early-life nutrition causes long-term alterations in DNA methylation, which have adverse effects on individual health. Nutrients can exert their influence by directly inhibiting epigenetic enzymes such as DNA methyl transferase (DNMT), Histone deacetylase (HDAC), or Histone acetyltransferases (HAT) or by changing the availability of the substances required for those enzymatic reactions. As a result, the expression of vital genes is modified, ultimately affecting overall well-being [28–30]. Studies have shown that nutrients have an epigenetic impact on phenotypic and disease susceptibility throughout life. Folate, a water-soluble B vitamin, is a source of one carbon to produce S-Adenosyl methionine (AdoMet), which is required for DNA methylation; folate metabolism is connected to phenotypic alterations through DNA methylation [29–31]. Other methyl donor substances, including choline, can likewise change the DNA's methylation status, which will then affect how genes are expressed [29]. A diet rich in methyl-donating nutrients can quickly affect gene expression, particularly in early development, when the epigenome is being established, and can have long term consequences in adulthood [32]. Animal studies have shown that an insufficient intake of folate or choline, which are methyl-donating compounds, before or after birth leads to lasting hypomethylation of certain genomic regions [33]. In adults, a diet lacking in a methyl group leads to reduced DNA methylation, but these changes can be reversed when methyl is reintroduced into the diet [32]. However, there is a gap in the literature in understanding the relationship between dietary intake of children and obesity-related gene methylation considering the racial disparities. In our previous study [34], we found that children who were overweight/obese (OW/OB) had increased methylation of the *NRF1* and *FTO* genes and decreased methylation of the *LEPR* gene when compared to normal weight (NW). Specifically, African American (AA) children had significantly higher methylation of *LEPR* compared to European American (EA) children. Along the line, this research aims to identify the dietary nutrients that are associated with the methylation of the obesity-related genes *NRF1*, *FTO*, and *LEPR* in racial disparities in childhood obesity.

2. Materials and Methods

2.1. Study Participants and Sample Analysis

Detailed information of the study participants has been previously given in Patel et al. [34]. Briefly, children aged 6 to 10 years were recruited from Lee and Macon counties, AL. A total of 113 children participated in this study. A phone survey was conducted with the parents to identify the children's history of diabetes or cardiovascular disease to exclude them from the study. Children who self-identify as EA or AA ethnicity were included. Children were brought by their parents to Auburn University, and their anthropometric data and saliva samples were collected. The body weights and heights of the participants were measured based on the World Health Organization (WHO)'s guidelines. The children were classified as normal weight, overweight, or obese, based on the Centers for Disease Control and Prevention (CDC)'s standards [35]. Furthermore, saliva was used to isolate the DNA, which was then bisulfite converted for the MethyLight RT-PCR reaction. The multiplex PCR was carried out for the three genes *NRF1*, *FTO*, and *LEPR*, for which two primers and one probe for each gene were designed. Detailed protocols for each step have been mentioned in previous published article [34].

2.2. Dietary Nutrient Analysis

Parents were asked to complete a detailed 24 h dietary recall of the child, which consisted of two parts. First, they were asked to write down what the child consumed for breakfast, lunch, and dinner, as well as desserts, snacks, and drinks. The second part consisted of the consumption of major food groups including vegetables, fruits, bread/grains, oily fish, high-fat meat, lean meat, non-meat protein, eggs, milk and milk products, cheese, sugared beverages, sweets, potato or corn snacks, and caffeinated beverages, along with their serving sizes. Additionally, each group had sub-categories of specific food items. For instance, the vegetable group had two sub-categories, uncooked and cooked vegetables, which had serving sizes of 1 tennis ball and $\frac{1}{2}$ a tennis ball, respectively. Similarly, for other groups, references were given for portion sizes such as 1 deck of cards, 1 golf ball, ounces, size of index finger, cubes, etc.

Using ESHA's Food Processor Diet Analysis Software Version 11.11 (Salem, OR, USA), energy and nutrient intake were calculated. The Food Processor Diet Analysis program from the ESHA offered thorough reports on nutritional consumption, both macro and micronutrients. Over 1900 food sources, including the USDA Standard Reference database, USDA Food Data Central Brands, and manufacturer's data, were included in the software database. Additionally, there were over 146,000 ingredients, recipes, and restaurant food brands in the software database.

2.3. Statistical Analysis

IBM SPSS Statistics 25.0 was used for all the statistical calculations. Based on the body weight, height, date of birth, and gender, the participants were divided into two groups: normal weight (NW) and overweight/obese (OW/OB). Similarly, participants were divided based on their racial groups: European American (EA) and African American (AA). To determine the differences in all the macro and micronutrient intake, an independent sample t-test was carried out between NW and OW/OB children as well as between EA and AA children. Pearson correlation coefficients were calculated to demonstrate the correlation between the DNA methylation of genes (*NRF1*, *FTO*, and *LEPR*) and each nutrient for the following categories: NW, OW/OB, EA, and AA. Additionally, the correlation was adjusted for variables, including maternal education, family income, gender, and age.

3. Results

The demographic details about the study population were given in the previous paper (1). Table 1 shows the PMRs of the *NRF1*, *FTO*, and *LEPR* genes and the nutrient intake by the children. In overweight/obese EA children, there was a notable rise in PMR (Percentage of Methylation Reference) for the *NRF1* and *FTO* genes, whereas no such increase was observed in the AA children. Conversely, the AA children had higher methylation levels of the *LEPR* gene among normal weight participants, but there were no differences in methylation between overweight/obese and normal weight EA children.

Table 1. General nutrient intake of children categorized by BMI and Race.

Nutrients	Normal Weight	Overweight/ Obese	*p* Value	European American	African American	*p* Value
PMR of *NRF1* (%)	68.925 ± 6.45	102.716 ± 13.031	**0.018**	75.692 ± 8.857	96.217 ± 11.592	0.155
PMR of *FTO* (%)	83.982 ± 17.499	168.24 ± 28.114	**0.010**	104.365 ± 19.412	147.612 ± 28.126	0.195
PMR of *LEPR* (%)	121.406 ± 13.544	85.078 ± 7.319	**0.025**	62.543 ± 2.307	157.065 ± 15.176	**0.000**
Calories (kcal)	2055.839 ± 110.41	2312.628 ± 118.573	0.116	2203.052 ± 113.764	2142.547 ± 116.239	0.714
Proteins (g)	75.416 ± 3.906	88.849 ± 6.974	0.086	83.291 ± 5.881	79.733 ± 4.852	0.653
Carbohydrates (g)	271.176 ± 16.539	303.724 ± 17.732	0.182	297.056 ± 17.906	273.068 ± 15.601	0.329
PUFA Fat (g)	9.187 ± 0.953	10.472 ± 1.356	0.432	8.861 ± 1.096	10.96 ± 1.198	0.200
Trans Fat (g)	0.214 ± 0.042	0.246 ± 0.056	0.647	0.216 ± 0.048	0.246 ± 0.049	0.657
Sugar (g)	109.305 ± 9.27	137.352 ± 12.289	0.067	130.186 ± 11.862	112.724 ± 8.678	0.259
Added Sugar (g)	38.791 ± 5.557	37.946 ± 7.282	0.926	34.928 ± 6.265	42.763 ± 6.403	0.389
Monosaccharide (g)	3.12 ± 0.63	3.122 ± 0.822	0.999	4.511 ± 0.812	1.37 ± 0.415	**0.002**
Disaccharide (g)	2.657 ± 0.558	2.867 ± 0.814	0.829	2.963 ± 0.513	2.493 ± 0.88	0.630
Oligosaccharides (g)	119.66 ± 6.833	158.091 ± 21.668	0.078	143.503 ± 18.634	130.355 ± 7.553	0.551
Vitamin A-IU (IU)	2723.929 ± 639.128	3116.007 ± 832.119	0.706	4133.752 ± 856.042	1363.155 ± 339.475	**0.007**
Vitamin A-RAE (mg)	424.487 ± 33.793	543.855 ± 39.448	**0.023**	479.89 ± 37.172	481.208 ± 36.895	0.980
Retinol (mcg)	334.447 ± 29.774	438.131 ± 33.648	**0.022**	350.436 ± 31.51	424.206 ± 32.143	0.108
Beta Carotene (mcg)	696.732 ± 130.705	830.128 ± 198.743	0.568	921.153 ± 178.052	555.361 ± 131.361	0.117
Vitamin B1 (mg)	1.095 ± 0.078	1.175 ± 0.101	0.527	1.043 ± 0.091	1.246 ± 0.082	0.108
Vitamin B2 (mg)	1.453 ± 0.105	1.634 ± 0.1	0.218	1.422 ± 0.092	1.685 ± 0.116	0.073
Vitamin B3 (mg)	15.682 ± 1.205	16.146 ± 1.437	0.804	13.595 ± 1.065	18.804 ± 1.518	**0.005**
Vitamin B3-NE (mg)	17.943 ± 1.267	17.947 ± 1.625	0.998	15.463 ± 1.075	21.07 ± 1.758	**0.005**
Vitamin B6 (mg)	1.315 ± 0.129	1.411 ± 0.141	0.613	1.147 ± 0.096	1.628 ± 0.17	**0.011**
Vitamin B12 (mcg)	3.336 ± 0.396	4.144 ± 0.424	0.166	3.088 ± 0.33	4.504 ± 0.49	**0.015**
Vitamin C (mg)	132.061 ± 32.341	151.565 ± 38.133	0.695	185.838 ± 43.068	84.976 ± 8.858	**0.042**
Vitamin D-IU (IU)	165.923 ± 17.027	216.319 ± 19.893	0.056	175.31 ± 17.498	207.703 ± 19.862	0.223
Folate (mcg)	267.393 ± 22.024	291.33 ± 30.814	0.522	263.297 ± 23.218	297.928 ± 30.041	0.356
Folate_DFE (mcg)	317.276 ± 35.577	373.59 ± 47.336	0.337	310.995 ± 35.098	384.883 ± 48.577	0.209
Pantothenic Acid (mg)	0.746 ± 0.092	0.691 ± 0.09	0.674	0.708 ± 0.091	0.736 ± 0.092	0.826
Fluoride (mg)	0.051 ± 0.034	0.012 ± 0.005	0.286	0.044 ± 0.033	0.019 ± 0.007	0.516
Iron (mg)	12.475 ± 0.739	14.885 ± 1.045	0.058	12.936 ± 0.861	14.449 ± 0.935	0.238
Manganese (mg)	0.566 ± 0.074	0.988 ± 0.485	0.364	0.867 ± 0.409	0.633 ± 0.087	0.615
Selenium (mcg)	45.071 ± 3.483	48.888 ± 5.108	0.530	44.638 ± 4.204	49.662 ± 4.311	0.411
Sodium (mg)	2799.205 ± 154.008	3618.817 ± 365.68	**0.033**	3291.305 ± 327.413	3047.947 ± 144.326	0.534
Omega 3 (g)	0.723 ± 0.078	0.745 ± 0.08	0.848	0.652 ± 0.066	0.835 ± 0.093	0.101
Omega 6 (g)	7.723 ± 0.815	7.214 ± 0.755	0.651	6.255 ± 0.609	9.07 ± 0.968	**0.012**

Data are expressed as mean ± SEM. *p* values are calculated with the *t*-test, which represents the statistical significance between NW and OW/OB participants and between EA and AA. Values in bold are statistically significant. *p* value was considered significant at 0.05 level.

OW/OB children had a significantly higher intake of vitamin A-RAE (mg) (543.855 ± 39.448, $p = 0.023$) and retinol (mcg) (438.131 ± 33.648, $p = 0.022$) compared to NW children. There was also a significantly higher consumption of sodium (mg) (3618.817 ± 365.68, $p = 0.03$) amongst the OW/OB children compared to the NW, whereas the EA children had higher monosaccharide intake (4.511 ± 0.812, $p = 0.002$), vitamin A (IU) intake (4133.752 ± 856.042, $p = 0.007$), and vitamin C (mg) intake (185.838 ± 43.068, $p = 0.04$) compared to the AA children. On the other hand, the AA children had a significantly higher intake of vitamin B3 (mg) (18.804 ± 1.518, $p = 0.005$), vitamin B6 (mg) (1.628 ± 0.17, $p = 0.011$), vitamin B12 (mcg) (4.504 ± 0.49, $p = 0.015$), and Omega 6 (g) (9.07 ± 0.968, $p = 0.012$) than the EA children.

Furthermore, the differences between the dietary intake of normal weight EA and AA, and overweight/obese EA and AA children were identified using an independent sample *t*-test, shown in Table 2. AA normal weight children had significantly higher intake of PUFA (g) (11.337 ± 1.71, $p = 0.028$), added sugars (g) (50.736 ± 9.33, $p = 0.036$), vitamin B1 (mg) (1.29 ± 0.117, $p = 0.014$), vitamin B2 (mg) (1.709 ± 0.161, $p = 0.017$), vitamin B3 (mg) (19.549 ± 2.008, $p = 0.001$), iron (mg) (13.996 ± 1.148, $p = 0.045$), manganese (mg) (0.729 ± 0.115, $p = 0.033$), omega 3 (g) (0.886 ± 0.135, $p = 0.042$), and omega 6 (g) (9.652 ± 1.402, $p = 0.021$) compared to EA normal weight. But, the EA normal weight children had a higher monosaccharide (g) intake (4.625 ± 1.019, $p = 0.012$) than the AA normal weight children. Conversely, among the overweight/obese EA and AA children, the EA children had a significantly higher sugar intake (160.285 ± 18.053, $p = 0.02$) and vitamin A intake (4442.696 ± 1329.737, $p = 0.048$), whereas the AA children had a higher intake of vitamin B12 (mcg) (5.278 ± 0.852, $p = 0.029$).

Table 2. Differences between the dietary intake of normal weight and overweight/obese EA and AA children.

Nutrients	Normal Weight			Overweight/Obese		
	European American	African American	*p* Value	European American	African American	*p* Value
PMR of *NRF1* (%)	48.869 ± 4.254	90.364 ± 11.351	**0.001**	101.676 ± 15.738	104.3 ± 23.001	0.923
PMR of *FTO* (%)	28.955 ± 1.965	142.805 ± 33.036	**0.001**	177.419 ± 33.618	154.252 ± 50.007	0.691
PMR of *LEPR* (%)	64.074 ± 3.23	182.692 ± 22.981	**0.000**	61.061 ± 3.322	121.675 ± 14.647	**0.000**
PUFA Fat (g)	7.176 ± 0.793	11.337 ± 1.71	**0.028**	10.493 ± 1.991	10.44 ± 1.643	0.985
Sugar (g)	99.117 ± 13.414	120.197 ± 12.663	0.259	160.285 ± 18.053	102.406 ± 10.96	**0.020**
Added Sugar (g)	27.617 ± 5.74	50.736 ± 9.33	**0.036**	42.01 ± 10.972	31.752 ± 7.773	0.496
Monosaccharide (g)	4.625 ± 1.019	1.511 ± 0.604	**0.012**	4.399 ± 1.273	1.175 ± 0.542	0.054
Vitamin A-IU (IU)	3814.841 ± 1088.595	1557.781 ± 574.091	0.077	4442.696 ± 1329.737	1094.385 ± 169.807	**0.048**
Vitamin B1 (mg)	0.913 ± 0.094	1.29 ± 0.117	**0.014**	1.169 ± 0.152	1.186 ± 0.112	0.936
Vitamin B2 (mg)	1.215 ± 0.124	1.709 ± 0.161	**0.017**	1.623 ± 0.126	1.652 ± 0.168	0.886
Vitamin B3 (mg)	12.065 ± 1.051	19.549 ± 2.008	**0.001**	15.077 ± 1.814	17.775 ± 2.358	0.363
Vitamin B3-NE (mg)	14.42 ± 1.017	21.708 ± 2.199	**0.003**	16.474 ± 1.875	20.19 ± 2.935	0.268
Vitamin B12 (mcg)	2.766 ± 0.544	3.944 ± 0.566	0.139	3.4 ± 0.382	5.278 ± 0.852	**0.029**
Iron (mg)	11.052 ± 0.887	13.996 ± 1.148	**0.045**	14.76 ± 1.401	15.076 ± 1.589	0.884
Manganese (mg)	0.414 ± 0.089	0.729 ± 0.115	**0.033**	1.307 ± 0.799	0.5 ± 0.13	0.421
Omega 3 (g)	0.571 ± 0.074	0.886 ± 0.135	**0.042**	0.731 ± 0.107	0.765 ± 0.121	0.839
Omega 6 (g)	5.919 ± 0.769	9.652 ± 1.402	**0.021**	6.58 ± 0.949	8.227 ± 1.241	0.293

Data are expressed as mean ± SEM. *p* values are calculated with the *t*-test, which represents the statistical significance between NW and OW/OB participants and between EA and AA. Values in bold are statistically significant. *p* value was considered significant at 0.05 level.

Additionally, it was important to understand the correlation of the methylation of the *NRF1*, *FTO*, and *LEPR* genes with individual nutrients for all the four groups: NW, OW/OB, EA, and AA. Table 3 shows the Pearson correlation between the *NRF1* methylation and the nutrient intake. The correlation between the *NRF1* methylation and the NW children nutrient intake demonstrated a significantly positive moderate correlation with added sugar ($r^2 = 0.296$, $p = 0.022$), oligosaccharides ($r^2 = 0.273$, $p = 0.035$), vitamin B2 ($r^2 = 0.360$, $p = 0.005$), vitamin B12 ($r^2 = 0.306$, $p = 0.017$), folate ($r^2 = 0.363$, $p = 0.004$), vitamin B6 ($r^2 = 0.385$, $p = 0.002$), and iron ($r^2 = 0.352$, $p = 0.006$). A stronger correlation was observed with Vitamin B1 ($r^2 = 0.455$, $p = 0.000$) and vitamin B3 ($r^2 = 0.431$, $p = 0.001$). On the other hand, a significant moderate correlation was observed between *NRF1* methylation and the OW/OB children's trans-fat ($r^2 = 0.339$, $p = 0.013$) intake and a stronger correlation with fluoride ($r^2 = 0.450$, $p = 0.001$). In all EA children, a positively moderate correlation was observed with iron intake ($r^2 = 0.264$, $p = 0.037$) and manganese ($r^2 = 0.358$, $p = 0.004$), whereas it was soluble fiber ($r^2 = 0.288$, $p = 0.043$) and pantothenic acid ($r^2 = 0.324$, $p = 0.022$)

for the AA children. A stronger correlation was seen with fluoride ($r^2 = 0.457$, $p = 0.001$) in the AA children. Furthermore, to understand the role of family income, maternal education, race, and gender, the adjusted person correlation was calculated for the *NRF1* methylation (Table 4). The results showed a positive correlation with total fiber ($r^2 = 0.267$, $p = 0.046$) intake in NW children after the adjusting, but the significance was lost for sugar, oligosaccharides, retinol, vitamin B2, vitamin B6, vitamin B12, folate, and manganese. If we look at the adjusted correlation in the OW/OB children, the significant correlation was lost for trans-fat. Among the races, the correlation was lost for iron in EA children.

Table 3. Pearson correlation coefficients (r^2) for the relation between DNA methylation of *NRF1* gene and nutrients.

Nutrients	Normal Weight		Overweight/Obese		European American		African American	
	r^2	*p* Value	r^2	*p* Value	r^2	*p* Value	r^2	*p* Value
Trans Fat (g)	−0.080	0.543	0.339	**0.013**	0.221	0.082	0.178	0.216
Fiber Soluble (g)	−0.090	0.496	0.155	0.268	0.156	0.222	0.288	**0.043**
Added Sugar (g)	0.296	**0.022**	−0.252	0.069	−0.102	0.426	−0.063	0.665
Oligosaccharide (g)	0.273	**0.035**	0.050	0.723	0.197	0.122	−0.007	0.963
Retinol (mcg)	0.301	**0.020**	−0.083	0.554	0.113	0.379	0.041	0.777
Vitamin B1 (mg)	0.455	**0.000**	−0.067	0.636	0.032	0.802	0.183	0.204
Vitamin B2 (mg)	0.360	**0.005**	−0.045	0.749	0.079	0.537	0.143	0.321
Vitamin B3 (mg)	0.431	**0.001**	−0.120	0.393	0.094	0.464	−0.008	0.958
Vitamin B3-NE (mg)	0.421	**0.001**	−0.136	0.333	0.024	0.850	0.003	0.982
Vitamin B6 (mg)	0.385	**0.002**	−0.127	0.367	0.011	0.932	0.056	0.700
Vitamin B12 (mcg)	0.306	**0.017**	−0.150	0.283	0.020	0.877	0.016	0.915
Folate (mcg)	0.363	**0.004**	−0.069	0.622	0.048	0.710	0.083	0.569
Folate_DFE (mcg)	0.283	**0.029**	−0.011	0.935	0.125	0.329	0.051	0.727
Pantothenic Acid (mg)	0.188	0.150	0.231	0.096	0.074	0.563	0.324	**0.022**
Fluoride (mg)	0.015	0.907	0.450	**0.001**	−0.015	0.908	0.457	**0.001**
Iron (mg)	0.352	**0.006**	0.037	0.791	0.264	**0.037**	0.034	0.814
Manganese (mg)	0.261	**0.044**	0.240	0.084	0.358	**0.004**	0.039	0.787

Values in bold are statistically significant. *p* value was considered significant at 0.05 level.

Table 5 shows the Pearson correlation between *FTO* methylation and the nutrient intake. The correlation between *FTO* methylation and NW children nutrient intake demonstrated a moderate positive correlation with calorie intake ($r^2 = 0.258$, $p = 0.046$), protein ($r^2 = 0.296$, $p = 0.022$), carbohydrate ($r^2 = 0.286$, $p = 0.026$), sugar ($r^2 = 0.264$, 0.042), PUFA ($r^2 = 0.276$, $p = 0.032$), vitamin B2 ($r^2 = 0.309$, $p = 0.016$), iron ($r^2 = 0.303$, $p = 0.019$), selenium ($r^2 = 0.298$, $p = 0.021$), omega 3 ($r^2 = 0.279$, $p = 0.031$), and omega 6 ($r^2 = 0.295$, $p = 0.022$). While nutrients such as vitamin B1 ($r^2 = 0.492$, $p = 0.00$) and vitamin B3 ($r^2 = 0.424$, $p = 0.001$) demonstrated a stronger correlation. Similar to *NRF1*, the *FTO* methylation in OW/OB children for trans-fat ($r^2 = 0.316$, $p = 0.021$) had a positive correlation. Along with that, disaccharide ($r^2 = 0.291$, $p = 0.035$), fluoride ($r^2 = 0.295$, $p = 0.034$), and manganese ($r^2 = 0.351$, $p = 0.010$) were significantly correlated. In children who were EA, nutrients such as vitamin D ($r^2 = 0.277$, $p = 0.029$), iron ($r^2 = 0.301$, $p = 0.017$), and manganese ($r^2 = 0.482$, $p = 0.00$) were positively associated. The intake of pantothenic acid ($r^2 = 0.332$, $p = 0.018$) was the only nutrient that demonstrated a moderate positive correlation with *FTO* methylation in the AA children. When adjusted for family income, maternal education, gender, and age (Table 6), the significant correlations for calories, sugar, PUFA, vitamin B2, and omega 3 intake were lost. For OW/OB children, the significance was not seen in the trans-fat and disaccharide intake. After adjusting the variables, vitamin D intake in EA and pantothenic acid intake in AA children were not correlated with *FTO* methylation.

Table 4. Adjusted Pearson correlation coefficients (r^2) for the relation between DNA methylation of *NRF1* gene and nutrients.

Nutrients	Normal Weight		Overweight/Obese		European American		African American	
	r^2	*p* Value	r^2	*p* Value	r^2	*p* Value	r^2	*p* Value
Fiber Total (g)	0.267	**0.046**	−0.010	0.948	0.092	0.487	0.067	0.656
Fiber Soluble Total (g)	0.099	0.470	0.185	0.204	0.253	0.053	0.343	**0.020**
Fiber Soluble (g)	0.052	0.705	0.169	0.245	0.242	0.065	0.363	**0.013**
Vitamin B1 (mg)	0.347	**0.009**	−0.011	0.942	0.004	0.974	0.194	0.195
Vitamin B3 (mg)	0.305	**0.022**	−0.070	0.632	0.094	0.477	0.000	0.999
Vitamin B3-NE (mg)	0.306	**0.022**	−0.085	0.562	0.030	0.819	0.009	0.955
Folate (mcg)	0.295	**0.027**	0.026	0.860	0.042	0.751	0.098	0.516
Pantothenic Acid (mg)	0.189	0.163	0.233	0.107	0.127	0.338	0.321	**0.030**
Fluoride (mg)	0.064	0.638	0.439	**0.002**	−0.055	0.683	0.392	**0.007**
Iron (mg)	0.281	**0.036**	0.089	0.543	0.244	0.063	0.060	0.692
Manganese (mg)	0.234	0.082	0.237	0.102	0.359	**0.005**	0.043	0.775

p values are adjusted for maternal education, family income, gender, and age. Values in bold are statistically significant. *p* value was considered significant at 0.05 level.

Table 5. Pearson correlation coefficients (r^2) for the relation between DNA methylation of *FTO* gene and nutrients.

Nutrients	Normal Weight		Overweight/Obese		European American		African American	
	r^2	*p* Value	r^2	*p* Value	r^2	*p* Value	r^2	*p* Value
Calories (kcal)	0.258	**0.046**	−0.189	0.175	0.057	0.658	0.027	0.851
Proteins (g)	0.296	**0.022**	−0.126	0.370	0.093	0.469	−0.005	0.974
Carbohydrates (g)	0.286	**0.026**	−0.232	0.094	−0.018	0.889	0.097	0.501
PUFA Fat (g)	0.276	**0.032**	−0.124	0.376	0.047	0.712	−0.001	0.993
Trans Fat (g)	−0.227	0.081	0.316	**0.021**	0.185	0.146	0.042	0.773
Sugar (g)	0.264	**0.042**	−0.170	0.222	0.052	0.685	0.050	0.730
Added Sugar (g)	0.323	**0.012**	−0.252	0.069	−0.095	0.461	0.006	0.969
Disaccharide (g)	−0.142	0.280	0.291	**0.035**	0.106	0.409	0.171	0.234
Vitamin B1 (mg)	0.492	**0.000**	−0.095	0.499	0.069	0.592	0.195	0.174
Vitamin B2 (mg)	0.309	**0.016**	−0.009	0.948	0.196	0.124	0.087	0.550
Vitamin B3 (mg)	0.424	**0.001**	−0.139	0.320	0.146	0.252	−0.008	0.954
Vitamin B3-NE (mg)	0.437	**0.000**	−0.181	0.195	0.064	0.618	−0.004	0.980
Vitamin D-IU (IU)	−0.015	0.911	0.035	0.801	0.277	**0.029**	−0.177	0.219
Pantothenic Acid (mg)	0.242	0.063	0.252	0.069	0.129	0.314	0.332	0.018
Fluoride (mg)	−0.043	0.742	0.295	**0.034**	−0.061	0.636	0.249	**0.081**
Iron (mg)	0.303	**0.019**	0.070	0.616	0.301	**0.017**	0.060	0.681
Manganese (mg)	−0.104	0.430	0.351	**0.010**	0.482	**0.000**	−0.156	0.278
Selenium (mcg)	0.298	**0.021**	−0.162	0.247	−0.002	0.988	0.013	0.928
Omega 3 (g)	0.279	**0.031**	−0.134	0.340	0.017	0.898	0.042	0.772
Omega 6 (g)	0.295	**0.022**	−0.063	0.659	0.084	0.511	0.040	0.782

Values in bold are statistically significant. *p* value was considered significant at 0.05 level.

Table 6. Adjusted Pearson correlation coefficients (r^2) for the relation between DNA methylation of *FTO* gene and nutrients.

Nutrients	Normal Weight		Overweight/Obese		European American		African American	
	r^2	*p* Value	r^2	*p* Value	r^2	*p* Value	r^2	*p* Value
Proteins (g)	0.286	**0.033**	−0.112	0.443	0.064	0.629	0.013	0.933
Carbohydrates (g)	0.302	**0.024**	−0.108	0.461	0.041	0.756	0.093	0.539
Vitamin B1 (mg)	0.456	**0.000**	−0.053	0.718	0.041	0.755	0.233	0.120
Vitamin B3 (mg)	0.364	**0.006**	−0.098	0.504	0.141	0.286	0.006	0.971
Vitamin B3-NE (mg)	0.387	**0.003**	−0.137	0.349	0.066	0.621	0.002	0.991
Fluoride (mg)	−0.012	0.931	0.295	**0.042**	−0.116	0.387	0.211	0.159
Iron (mg)	0.248	0.066	0.170	0.244	0.284	**0.029**	0.128	0.395
Manganese (mg)	−0.155	0.255	0.336	**0.018**	0.495	**0.000**	−0.156	0.300
Selenium (mcg)	0.303	**0.023**	−0.193	0.183	−0.052	0.696	0.019	0.903
Omega 6 (g)	0.267	**0.046**	−0.118	0.425	0.036	0.787	0.018	0.908

p values are adjusted for maternal education, family income, gender, and age. Values in bold are statistically significant. *p* value was considered significant at 0.05 level.

Table 7 shows the Pearson correlation between *LEPR* methylation and the nutrient intake. DNA methylation of the gene *LEPR* was moderately correlated with only one nutrient: manganese intake in NW children ($r^2 = 0.336$, $p = 0.009$). OW/OB children's intake of trans-fat ($r^2 = 0.361$, $p = 0.008$) and fluoride ($r^2 = 0.477$, $p = 0.000$) were strongly correlated. While, for EA children, added sugar ($r^2 = -0.253$, $p = 0.045$), monosaccharide ($r^2 = -0.287$, $p = 0.023$), and beta carotene ($r^2 = -0.312$, $p = 0.013$) had moderate negative correlations. Similar to the OW/OB children, AA children's intake of trans-fat intake ($r^2 = 0.294$, $p = 0.038$) was correlated with *LEPR* methylation. After Pearson correlation, adjusting for family income, maternal education, race, and gender (Table 8), significance was lost for monosaccharide and added sugar among EA children and for trans-fat and pantothenic acid among AA children. This suggested that the relationship between methylation and the dietary nutrient intake may have been dependent upon the races, genders, family incomes, and maternal educations of the individuals.

Table 7. Pearson correlation coefficients (r^2) for the relation between DNA methylation of *LEPR* gene and nutrients.

Nutrients	Normal Weight		Overweight/Obese		European American		African American	
	r^2	*p* Value	r^2	*p* Value	r^2	*p* Value	r^2	*p* Value
Trans Fat (g)	0.141	0.282	0.361	**0.008**	0.089	0.487	0.294	**0.038**
Added Sugar (g)	0.102	0.437	−0.229	0.099	−0.253	**0.045**	−0.050	0.732
Monosaccharide (g)	−0.121	0.357	−0.264	0.056	−0.287	**0.023**	0.151	0.295
Beta Carotene (mcg)	−0.173	0.186	−0.270	0.051	−0.312	**0.013**	−0.172	0.233
Pantothenic Acid (mg)	0.156	0.232	0.252	0.068	0.202	0.112	0.284	**0.046**
Fluoride (mg)	−0.047	0.723	0.477	**0.000**	0.241	0.059	0.017	0.908
Manganese (mg)	0.336	**0.009**	0.078	0.577	0.376	**0.002**	0.238	0.096

Values in bold are statistically significant. *p* value was considered significant at 0.05 level.

Table 8. Adjusted Pearson correlation coefficients (r^2) for the relation between DNA methylation of *LEPR* gene and nutrients.

Nutrients	Normal Weight		Overweight/Obese		European American		African American	
	r^2	*p* Value	r^2	*p* Value	r^2	*p* Value	r^2	*p* Value
Trans Fat (g)	0.151	0.266	0.326	**0.022**	0.075	0.572	0.195	0.194
Beta Carotene (mcg)	−0.212	0.117	−0.247	0.088	−0.359	**0.005**	−0.173	0.250
Fluoride (mg)	−0.017	0.899	0.468	**0.001**	0.218	0.100	0.022	0.884
Manganese (mg)	0.296	**0.027**	0.082	0.575	0.371	**0.004**	0.206	0.170

p values are adjusted for maternal education, family income, gender, and age. Values in bold are statistically significant. *p* value was considered significant at 0.05 level.

Lastly, Figure 1 shows the biochemical pathways involving various nutrients in the one carbon metabolism and the link towards the methylation of the *NRF1*, *FTO*, and *LEPR* genes.

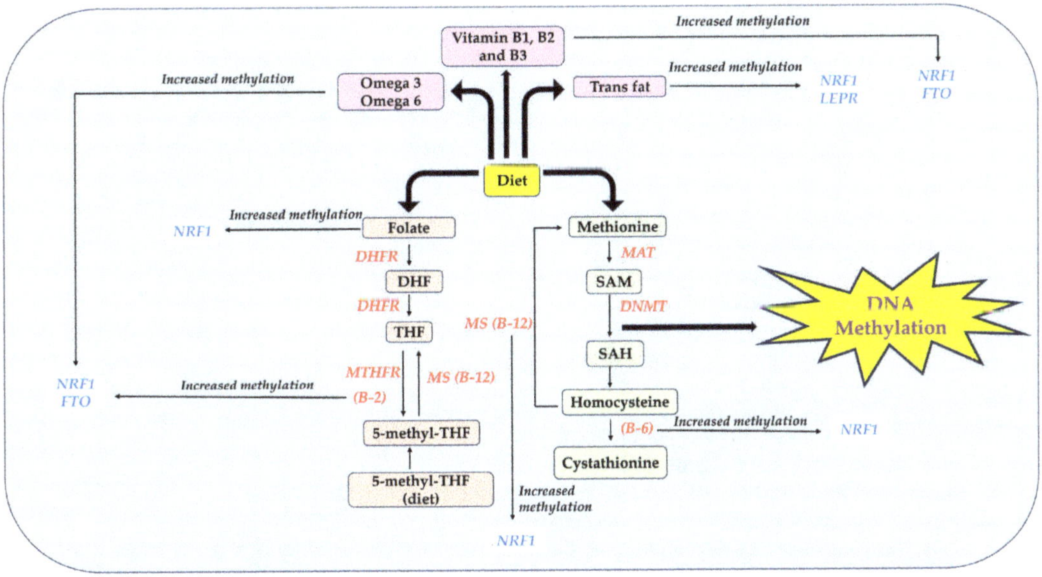

Figure 1. Summary of nutrients involved in one-carbon metabolism and the effects on the methylation of *NRF1*, *FTO*, and *LEPR* genes. B2, riboflavin; B3, niacin; B6, pyridoxine; B12, vitamin B12; B1, thiamine; NRF1, Nuclear Respiratory Factor 1; FTO, Fat mass and obesity associated; LEPR, Leptin Receptor; DHF, Dihydrofolic acid; THF, Tetrahydrofolate; SAM, S-adenosylmethionine; SAH, S-adenosylhomocysteine.

4. Discussion

The study explored the association between the DNA methylation of the three genes *NRF1*, *FTO*, and *LEPR* with the nutrient intake of children. DNA methylation is a well-studied epigenetic modification that takes place in the one-carbon metabolism pathway. This process relies on specific enzymes and dietary micronutrients such as folate, choline, and betaine [36]. In this pathway, methionine is converted to S-adenosylmethionine (SAM), which serves as a methyl donor in cells. DNA methyltransferases (DNMTs) use SAM to attach methyl groups to the carbon-5 position of cytosine bases in DNA, resulting in the methylation of DNA [37].

Under normal conditions, folate that is consumed through a diet undergoes a process of metabolism in the intestine and/or liver, resulting in the formation of 5-methyltetrahydro folate (5-methylTHF) in its monoglutamyl form, which further needs to be converted to tetrahydrofolate (THF) [38]. Polyglutamate synthetase is most effective when using THF as a substrate; thus, 5-methylTHF must be converted to THF through the methionine synthase reaction. Once THF is formed, either from folic acid or dietary folate, it is initially transformed into 5,10-methyleneTHF by the vitamin B6-dependent enzyme serine hydroxy methyltransferase. It is then irreversibly reduced to 5-methylTHF by methylenetetrahydrofolate reductase (MTHFR) [38]. This conversion is crucial for maintaining a steady supply of methyl groups used in the remethylation of homocysteine to methionine, which is facilitated by the vitamin B12-dependent enzyme methionine synthase [38]. Methionine serves as a substrate for SAM, a cofactor and methyl group donor involved in various methylation reactions, including the methylation of DNA, RNA, neurotransmitters, and other small molecules [39,40]. In this study, the dietary folate intake of NW children was associated with the DNA methylation of the *NRF1* gene. We have previously mentioned that the methylation level of *NRF1* in NW children was significantly lower than in OW/OB children [34]: the dietary intake of folate is lower in NW but not significantly different compared to OW/OB children. Ramos-Lopez and colleagues [41] conducted a cross-sectional study to examine how folate intake is related to the genomic methylation profile in a group of 47 obese participants from the Metabolic Syndrome Reduction in Navarra-Spain trial. They discovered that there were 51 CpGs (regions of DNA) that showed an association with folate intake. One of our genes of interest, *NRF1*, was associated with the innate immune response and plays a role in regulating various aspects of brown adipose tissue, including thermogenic adaptation, adipocyte inflammation, and cytokine production [42,43]. On the other hand, the gene *FTO* is responsible for controlling energy balance and eating behavior in specific regions of the hypothalamus, such as the arcuate, paraventricular, dorsomedial, and ventromedial nuclei [44]. Studies conducted in both living organisms and laboratory settings have shown that *FTO* can detect the nutritional status of the body and respond to appetite and food intake [45].

Several previous studies have examined the relationships between BMI, fat mass, and folate concentration in various age groups and populations [46–49]. One study found two possible explanations for the associations observed: firstly, obesity may lead to low serum folate levels, potentially affecting how the body processes folate and increasing the need for dietary folate. Secondly, low serum folate levels could be a contributing factor to obesity by affecting epigenetic modifications involved in lipid metabolism [50]. Another study supported these findings by showing that individuals with a BMI greater than 25 kg/m^2 had lower serum folate levels, regardless of their folate intake [51]. This supported the hypothesis proposed by da Silva et al. that said that obesity may independently impact folate distribution by increasing the cellular uptake of dietary folate [52]. The findings of the above-mentioned studies could support our results of the OW/OB children having a higher methylation level for the *NRF1* gene, irrespective of their folate intake. It is interesting to note that the association was observed to be weaker after adjusting for the covariates of the study. Additionally, if we look at the methylation difference for *NRF1* in EA and AA normal weight children, there was a significant increase in the methylation of AA normal weight children, which could explain the positive association between *NRF1* methylation and folate intake. Furthermore, it would be interesting to see if the reduction in folate intake for OW/OB children could reduce the methylation of the genes.

Other methyl donors from the diet are vitamins B2 and B6. The riboflavin intake in the NW children of this study had a positive association with the methylation of the *NRF1* and *FTO* genes. In addition, the intake of riboflavin and the DNA methylation of the *NRF1* and *FTO* were significantly higher in OW/OB children. Similarly, significantly higher vitamin B6 intake and *NRF1* methylation were observed in AA children. Even though not significant, OW/OB children had a higher intake than the NW children. Previous research has shown that a higher intake of methyl donors could increase the methylation levels of the gene [53,54]. Even though not significantly associated with OW/OB children, normal-weight AA children had significantly higher vitamin B2 intake. Additionally, after adjusting for the covariates, the significant association was lost, indicating the impact of socio-economic status (maternal education, family income) gender, and age in the association.

The negative impacts of trans fatty acids in our diet and their effects on human health have been extensively studied and proven. Despite being restricted or prohibited in many countries, trans fatty acids might still lead to prolonged reactions that could raise valid concerns about human health, especially if they modify the epigenome [55]. Our study showed a positive correlation between *NRF1*, *FTO*, and *LEPR* methylation and OW/OB children's trans-fat intake. Studies have been carried out to understand the role of fat intake on the epigenome [55–58], one of which suggested that the 54 genes associated exclusively with SFA CpGs have a significant impact on the liver's metabolic functions, particularly in the regulation of glucose and insulin metabolism in obese adults [59]. Additionally, the study observed a higher correlation between DNA methylation of the CpGs associated with SFA and the presence of palmitic and stearic acid [59]. Unfortunately, we do not have human studies to support the association between trans-fat and DNA methylation.

Sufficient intake of micronutrients, including manganese (Mn), is crucial for proper fetal development. Imbalances in Mn levels, both deficiencies and excessive exposure, have been linked to the development of diseases later in life [60]. Bozack et al. examined the relationship between Mn levels in maternal erythrocytes during the first trimester and the presence of differentially methylated positions and regions in cord blood [60]. They also investigated whether these associations persisted in blood samples collected from children at mid-childhood (6–10 years old) in a cohort of 361 individuals. The study revealed that Mn levels were associated with increased methylation of a specific DNA site, cg02042823, located in the gene RNA binding fox-1 homolog 1 (RBFOX1 or A2BP1), in cord blood. This association remained significant but was weakened in blood samples collected during mid-childhood. The findings suggested a connection between prenatal levels of micronutrients, epigenetic modifications in the placenta, and birth weight [61]. The influence of Mn on epigenetic processes is an emerging field of research, and, thus far, only one human study has reported Mn-related changes in DNA methylation from birth to childhood. Our results showed increased methylations for all the three genes along with a significant correlation with the EA children manganese intake, suggesting to further conduct more race-specific human research for manganese and DNA methylation.

To our knowledge, this is the first study that demonstrated the association of the dietary intake of children with the methylation of obesity-related genes, considering the racial and health disparity effects. The prevalence rates of obesity varied significantly based on race and ethnicity. African Americans had a 50% higher likelihood of being obese compared to non-Hispanic whites [62]. These findings were consistent with another study that confirmed the higher risk of obesity among African Americans [63].

It has been observed that race, along with BMI status, is associated with the DNA methylation of genes, suggesting that epigenetic regulation may contribute to health disparities among different racial and ethnic groups [64]. Some significant changes were seen after adjusting the correlation in this study, implying the role of socioeconomic status in food availability, food security, and even the perception of healthy food. There is a growing body of evidence [65–69] supporting the connection between obesity and food insecurity, particularly for women, although the findings for children are still mixed. A study [70] revealed that both adults and children had high rates of overweight and obesity, and there was a significant prevalence of families that had recently experienced food insecurity. Living in a household that is food secure was linked to the perception that healthy food options are both affordable and convenient. Caregivers from food-insecure households experiencing hunger were found to have higher rates of unemployment and lower incomes compared to those from food-secure households [70]. Even though there were not many significant differences in the nutrient intake of OW/OB children, the lack of a significant correlation between nutrient intake and methylation of genes suggested the importance of environmental factors. The results also demonstrated the possible increased risk of normal-weight AA children becoming obese in the future. Along with this, it also opens the window to reevaluate the dietary needs for each nutrient based on their racial and socioeconomic status.

While our research established a connection between dietary nutrients and DNA methylation, as well as a potential association with health disparities, the limited size of our study sample might have hindered the identification of significant correlations with obesity indicators. To enhance the reliability and validity of the findings, it would be beneficial to broaden the scope of the current investigation and include larger sample sizes. Additionally, the diet intake of children was self-reported by the mothers, who could have under or overestimated certain portion sizes, affecting the overall dietary assessment.

5. Conclusions

To conclude, we found a significant association between the nutrient intake of children and the methylation of the *NRF1*, *FTO*, and *LEPR* genes. The study also highlighted the importance of health disparities in understanding the epigenetic effects of methyl donors in the diet and the methylation of genes. Furthermore, intervention studies can help observe the role of nutrient intake, specifically methyl donors in the diet in assessing the risk factors for childhood obesity.

Author Contributions: Conceptualization, J.R.B. and T.G.; data curation, P.P. and V.S.; formal analysis, P.P. and V.S.; funding acquisition, J.R.B. and T.G.; methodology, P.P. and V.S.; project administration, T.G.; supervision, T.G.; visualization, J.R.B. and T.G.; writing—original draft preparation P.P.; writing—review and editing, all. All authors have read and agreed to the published version of the manuscript.

Funding: This work was supported by the Alabama Agricultural Experiment Station and the Hatch Program of the National Institute of Food and Agriculture, US Department of Agriculture, to J.R.B. and T.G.

Institutional Review Board Statement: This study was conducted according to the guidelines of the Declaration of Helsinki and approved by the Institutional Review Board of Auburn University (Protocol # 17-364 MR 1709).

Informed Consent Statement: Informed consent was obtained from all participants involved in the study.

Data Availability Statement: The study datasets of the current manuscript are available from the corresponding author upon request.

Acknowledgments: We would like to thank all the children for their participation.

Conflicts of Interest: The authors declare no conflict of interest.

References

1. Rhee, K.E.; Phelan, S.; McCaffery, J. Early determinants of obesity: Genetic, epigenetic, and in utero influences. *Int. J. Pediatr.* **2012**, *2012*, 463850. [CrossRef] [PubMed]
2. NCD Risk Factor Collaboration (NCD-RisC). Worldwide trends in body-mass index, underweight, overweight, and obesity from 1975 to 2016: A pooled analysis of 2416 population-based measurement studies in 128.9 million children, adolescents, and adults. *Lancet* **2017**, *390*, 2627–2642. [CrossRef] [PubMed]
3. Qasim, A.; Turcotte, M.; de Souza, R.J.; Samaan, M.C.; Champredon, D.; Dushoff, J.; Speakman, J.R.; Meyre, D. On the origin of obesity: Identifying the biological, environmental and cultural drivers of genetic risk among human populations. *Obes. Rev.* **2018**, *19*, 121–149. [CrossRef] [PubMed]
4. Pantalone, K.M.; Hobbs, T.M.; Chagin, K.M.; Kong, S.X.; Wells, B.J.; Kattan, M.W.; Bouchard, J.; Sakurada, B.; Milinovich, A.; Weng, W.; et al. Prevalence and recognition of obesity and its associated comorbidities: Cross-sectional analysis of electronic health record data from a large US integrated health system. *BMJ Open* **2017**, *7*, e017583. [CrossRef]
5. Smith, K.B.; Smith, M.S. Obesity Statistics. *Prim. Care* **2016**, *43*, 121–135. [CrossRef]
6. Simmonds, M.; Llewellyn, A.; Owen, C.G.; Woolacott, N. Predicting adult obesity from childhood obesity: A systematic review and meta-analysis. *Obes. Rev.* **2016**, *17*, 95–107. [CrossRef]
7. Johnson, V.R.; Acholonu, N.O.; Dolan, A.C.; Krishnan, A.; Wang, E.H.; Stanford, F.C. Racial Disparities in Obesity Treatment Among Children and Adolescents. *Curr. Obes. Rep.* **2021**, *10*, 342–350. [CrossRef]

8. Weinhold, B. Epigenetics: The science of change. *Environ. Health Perspect.* **2006**, *114*, A160–A167. [CrossRef]
9. Pokrywka, M.; Kiec-Wilk, B.; Polus, A.; Wybranska, I. DNA methylation in obesity. *Postepy Hig. Med. Dosw. (Online)* **2014**, *68*, 1383–1391. [CrossRef]
10. Liu, L.; Li, Y.; Tollefsbol, T.O. Gene-environment interactions and epigenetic basis of human diseases. *Curr. Issues Mol. Biol.* **2008**, *10*, 25–36.
11. Giglio, R.V.; Stoian, A.P.; Patti, A.M.; Rizvi, A.A.; Sukhorukov, V.; Ciaccio, M.; Orekhov, A.; Rizzo, M. Genetic and Epigenetic Biomarkers for Diagnosis, Prognosis and Treatment of Metabolic Syndrome. *Curr. Pharm. Des.* **2021**, *27*, 3729–3740. [CrossRef]
12. Sayols-Baixeras, S.; Subirana, I.; Fernandez-Sanles, A.; Senti, M.; Lluis-Ganella, C.; Marrugat, J.; Elosua, R. DNA methylation and obesity traits: An epigenome-wide association study. The REGICOR study. *Epigenetics* **2017**, *12*, 909–916. [CrossRef]
13. Crujeiras, A.B.; Diaz-Lagares, A.; Sandoval, J.; Milagro, F.I.; Navas-Carretero, S.; Carreira, M.C.; Gomez, A.; Hervas, D.; Monteiro, M.P.; Casanueva, F.F.; et al. DNA methylation map in circulating leukocytes mirrors subcutaneous adipose tissue methylation pattern: A genome-wide analysis from non-obese and obese patients. *Sci. Rep.* **2017**, *7*, 41903. [CrossRef]
14. Samblas, M.; Milagro, F.I.; Martinez, A. DNA methylation markers in obesity, metabolic syndrome, and weight loss. *Epigenetics* **2019**, *14*, 421–444. [CrossRef]
15. Shen, J.; Zhu, B. Integrated analysis of the gene expression profile and DNA methylation profile of obese patients with type 2 diabetes. *Mol. Med. Rep.* **2018**, *17*, 7636–7644. [CrossRef]
16. Castellano-Castillo, D.; Moreno-Indias, I.; Sanchez-Alcoholado, L.; Ramos-Molina, B.; Alcaide-Torres, J.; Morcillo, S.; Ocana-Wilhelmi, L.; Tinahones, F.; Queipo-Ortuno, M.I.; Cardona, F. Altered Adipose Tissue DNA Methylation Status in Metabolic Syndrome: Relationships Between Global DNA Methylation and Specific Methylation at Adipogenic, Lipid Metabolism and Inflammatory Candidate Genes and Metabolic Variables. *J. Clin. Med.* **2019**, *8*, 87. [CrossRef]
17. Crujeiras, A.B.; Diaz-Lagares, A.; Moreno-Navarrete, J.M.; Sandoval, J.; Hervas, D.; Gomez, A.; Ricart, W.; Casanueva, F.F.; Esteller, M.; Fernandez-Real, J.M. Genome-wide DNA methylation pattern in visceral adipose tissue differentiates insulin-resistant from insulin-sensitive obese subjects. *Transl. Res.* **2016**, *178*, 13–24.e5. [CrossRef]
18. Lorenzo, P.M.; Izquierdo, A.G.; Diaz-Lagares, A.; Carreira, M.C.; Macias-Gonzalez, M.; Sandoval, J.; Cueva, J.; Lopez-Lopez, R.; Casanueva, F.F.; Crujeiras, A.B. ZNF577 Methylation Levels in Leukocytes From Women With Breast Cancer Is Modulated by Adiposity, Menopausal State, and the Mediterranean Diet. *Front. Endocrinol.* **2020**, *11*, 245. [CrossRef]
19. Waddington, C.H. The epigenotype. 1942. *Int. J. Epidemiol.* **2012**, *41*, 10–13. [CrossRef]
20. Milagro, F.I.; Campion, J.; Cordero, P.; Goyenechea, E.; Gomez-Uriz, A.M.; Abete, I.; Zulet, M.A.; Martinez, J.A. A dual epigenomic approach for the search of obesity biomarkers: DNA methylation in relation to diet-induced weight loss. *FASEB J.* **2011**, *25*, 1378–1389. [CrossRef]
21. Moleres, A.; Campion, J.; Milagro, F.I.; Marcos, A.; Campoy, C.; Garagorri, J.M.; Gomez-Martinez, S.; Martinez, J.A.; Azcona-Sanjulian, M.C.; Marti, A.; et al. Differential DNA methylation patterns between high and low responders to a weight loss intervention in overweight or obese adolescents: The EVASYON study. *FASEB J.* **2013**, *27*, 2504–2512. [CrossRef] [PubMed]
22. Izquierdo, A.G.; Crujeiras, A.B. Obesity-Related Epigenetic Changes After Bariatric Surgery. *Front. Endocrinol.* **2019**, *10*, 232. [CrossRef] [PubMed]
23. Nicoletti, C.F.; Pinhel, M.A.S.; Diaz-Lagares, A.; Casanueva, F.F.; Jacome, A.; Pinhanelli, V.C.; de Oliveira, B.A.P.; Crujeiras, A.B.; Nonino, C.B. DNA methylation screening after roux-en Y gastric bypass reveals the epigenetic signature stems from genes related to the surgery per se. *BMC Med. Genom.* **2019**, *12*, 72. [CrossRef] [PubMed]
24. Nitert, M.D.; Dayeh, T.; Volkov, P.; Elgzyri, T.; Hall, E.; Nilsson, E.; Yang, B.T.; Lang, S.; Parikh, H.; Wessman, Y.; et al. Impact of an exercise intervention on DNA methylation in skeletal muscle from first-degree relatives of patients with type 2 diabetes. *Diabetes* **2012**, *61*, 3322–3332. [CrossRef]
25. Ronn, T.; Volkov, P.; Davegardh, C.; Dayeh, T.; Hall, E.; Olsson, A.H.; Nilsson, E.; Tornberg, A.; Dekker Nitert, M.; Eriksson, K.F.; et al. A six months exercise intervention influences the genome-wide DNA methylation pattern in human adipose tissue. *PLoS Genet.* **2013**, *9*, e1003572. [CrossRef]
26. Arora, I.; Sharma, M.; Sun, L.Y.; Tollefsbol, T.O. The Epigenetic Link between Polyphenols, Aging and Age-Related Diseases. *Genes* **2020**, *11*, 1094. [CrossRef]
27. Lorenzo, P.M.; Izquierdo, A.G.; Rodriguez-Carnero, G.; Fernandez-Pombo, A.; Iglesias, A.; Carreira, M.C.; Tejera, C.; Bellido, D.; Martinez-Olmos, M.A.; Leis, R.; et al. Epigenetic Effects of Healthy Foods and Lifestyle Habits from the Southern European Atlantic Diet Pattern: A Narrative Review. *Adv. Nutr.* **2022**, *13*, 1725–1747. [CrossRef]
28. Lillycrop, K.A.; Hoile, S.P.; Grenfell, L.; Burdge, G.C. DNA methylation, ageing and the influence of early life nutrition. *Proc. Nutr. Soc.* **2014**, *73*, 413–421. [CrossRef]
29. Choi, S.W.; Friso, S. Epigenetics: A New Bridge between Nutrition and Health. *Adv. Nutr.* **2010**, *1*, 8–16. [CrossRef]
30. Davis, C.D.; Ross, S.A. Dietary components impact histone modifications and cancer risk. *Nutr. Rev.* **2007**, *65*, 88–94. [CrossRef]
31. Kim, K.C.; Friso, S.; Choi, S.W. DNA methylation, an epigenetic mechanism connecting folate to healthy embryonic development and aging. *J. Nutr. Biochem.* **2009**, *20*, 917–926. [CrossRef]
32. Alavian-Ghavanini, A.; Ruegg, J. Understanding Epigenetic Effects of Endocrine Disrupting Chemicals: From Mechanisms to Novel Test Methods. *Basic Clin. Pharmacol. Toxicol.* **2018**, *122*, 38–45. [CrossRef]
33. Bokor, S.; Vass, R.A.; Funke, S.; Ertl, T.; Molnar, D. Epigenetic Effect of Maternal Methyl-Group Donor Intake on Offspring's Health and Disease. *Life* **2022**, *12*, 609. [CrossRef]

34. Patel, P.; Selvaraju, V.; Babu, J.R.; Wang, X.; Geetha, T. Racial Disparities in Methylation of NRF1, FTO, and LEPR Gene in Childhood Obesity. *Genes* **2022**, *13*, 2030. [CrossRef]
35. Kuczmarski, R.J.; Ogden, C.L.; Guo, S.S.; Grummer-Strawn, L.M.; Flegal, K.M.; Mei, Z.; Wei, R.; Curtin, L.R.; Roche, A.F.; Johnson, C.L. *2000 CDC Growth Charts for the United States: Methods and Development*; Department of Health and Human Services, Centers for Disease Control and Prevention, National Center for Health Statistics: Atlanta, GA, USA, 2002; pp. 1–190.
36. Anderson, O.S.; Sant, K.E.; Dolinoy, D.C. Nutrition and epigenetics: An interplay of dietary methyl donors, one-carbon metabolism and DNA methylation. *J. Nutr. Biochem.* **2012**, *23*, 853–859. [CrossRef]
37. Selhub, J. Homocysteine metabolism. *Annu. Rev. Nutr.* **1999**, *19*, 217–246. [CrossRef]
38. Crider, K.S.; Yang, T.P.; Berry, R.J.; Bailey, L.B. Folate and DNA methylation: A review of molecular mechanisms and the evidence for folate's role. *Adv. Nutr.* **2012**, *3*, 21–38. [CrossRef]
39. Bailey, R.L.; Mills, J.L.; Yetley, E.A.; Gahche, J.J.; Pfeiffer, C.M.; Dwyer, J.T.; Dodd, K.W.; Sempos, C.T.; Betz, J.M.; Picciano, M.F. Unmetabolized serum folic acid and its relation to folic acid intake from diet and supplements in a nationally representative sample of adults aged > or =60 y in the United States. *Am. J. Clin. Nutr.* **2010**, *92*, 383–389. [CrossRef]
40. Stover, P.J. One-carbon metabolism-genome interactions in folate-associated pathologies. *J. Nutr.* **2009**, *139*, 2402–2405. [CrossRef]
41. Ramos-Lopez, O.; Samblas, M.; Milagro, F.I.; Zulet, M.A.; Mansego, M.L.; Riezu-Boj, J.I.; Martinez, J.A. Association of low dietary folate intake with lower CAMKK2 gene methylation, adiposity, and insulin resistance in obese subjects. *Nutr. Res.* **2018**, *50*, 53–62. [CrossRef]
42. Bartelt, A.; Widenmaier, S.B.; Schlein, C.; Johann, K.; Goncalves, R.L.S.; Eguchi, K.; Fischer, A.W.; Parlakgul, G.; Snyder, N.A.; Nguyen, T.B.; et al. Brown adipose tissue thermogenic adaptation requires Nrf1-mediated proteasomal activity. *Nat. Med.* **2018**, *24*, 292–303. [CrossRef] [PubMed]
43. Wang, W.; Chan, J.Y. Nrf1 is targeted to the endoplasmic reticulum membrane by an N-terminal transmembrane domain. Inhibition of nuclear translocation and transacting function. *J. Biol. Chem.* **2006**, *281*, 19676–19687. [CrossRef] [PubMed]
44. Zhao, X.; Yang, Y.; Sun, B.F.; Zhao, Y.L.; Yang, Y.G. FTO and obesity: Mechanisms of association. *Curr. Diabetes Rep.* **2014**, *14*, 486. [CrossRef] [PubMed]
45. Dina, C.; Meyre, D.; Gallina, S.; Durand, E.; Korner, A.; Jacobson, P.; Carlsson, L.M.; Kiess, W.; Vatin, V.; Lecoeur, C.; et al. Variation in FTO contributes to childhood obesity and severe adult obesity. *Nat. Genet.* **2007**, *39*, 724–726. [CrossRef]
46. Gunanti, I.R.; Marks, G.C.; Al-Mamun, A.; Long, K.Z. Low serum vitamin B-12 and folate concentrations and low thiamin and riboflavin intakes are inversely associated with greater adiposity in Mexican American children. *J. Nutr.* **2014**, *144*, 2027–2033. [CrossRef]
47. Mahabir, S.; Ettinger, S.; Johnson, L.; Baer, D.J.; Clevidence, B.A.; Hartman, T.J.; Taylor, P.R. Measures of adiposity and body fat distribution in relation to serum folate levels in postmenopausal women in a feeding study. *Eur. J. Clin. Nutr.* **2008**, *62*, 644–650. [CrossRef]
48. Kimmons, J.E.; Blanck, H.M.; Tohill, B.C.; Zhang, J.; Khan, L.K. Associations between body mass index and the prevalence of low micronutrient levels among US adults. *MedGenMed* **2006**, *8*, 59.
49. Tinker, S.C.; Hamner, H.C.; Berry, R.J.; Bailey, L.B.; Pfeiffer, C.M. Does obesity modify the association of supplemental folic acid with folate status among nonpregnant women of childbearing age in the United States? *Birth Defects Res. A Clin. Mol. Teratol.* **2012**, *94*, 749–755. [CrossRef]
50. Mlodzik-Czyzewska, M.A.; Malinowska, A.M.; Chmurzynska, A. Low folate intake and serum levels are associated with higher body mass index and abdominal fat accumulation: A case control study. *Nutr. J.* **2020**, *19*, 53. [CrossRef]
51. Oliai Araghi, S.; Braun, K.V.E.; van der Velde, N.; van Dijk, S.C.; van Schoor, N.M.; Zillikens, M.C.; de Groot, L.; Uitterlinden, A.G.; Stricker, B.H.; Voortman, T.; et al. B-vitamins and body composition: Integrating observational and experimental evidence from the B-PROOF study. *Eur. J. Nutr.* **2020**, *59*, 1253–1262. [CrossRef]
52. da Silva, V.R.; Hausman, D.B.; Kauwell, G.P.; Sokolow, A.; Tackett, R.L.; Rathbun, S.L.; Bailey, L.B. Obesity affects short-term folate pharmacokinetics in women of childbearing age. *Int. J. Obes.* **2013**, *37*, 1608–1610. [CrossRef]
53. Mahmoud, A.M.; Ali, M.M. Methyl Donor Micronutrients that Modify DNA Methylation and Cancer Outcome. *Nutrients* **2019**, *11*, 608. [CrossRef]
54. Shorter, K.R.; Felder, M.R.; Vrana, P.B. Consequences of dietary methyl donor supplements: Is more always better? *Prog. Biophys. Mol. Biol.* **2015**, *118*, 14–20. [CrossRef]
55. Flores-Sierra, J.; Arredondo-Guerrero, M.; Cervantes-Paz, B.; Rodriguez-Rios, D.; Alvarado-Caudillo, Y.; Nielsen, F.C.; Wrobel, K.; Wrobel, K.; Zaina, S.; Lund, G. The trans fatty acid elaidate affects the global DNA methylation profile of cultured cells and in vivo. *Lipids Health Dis.* **2016**, *15*, 75. [CrossRef]
56. Lai, C.Q.; Parnell, L.D.; Smith, C.E.; Guo, T.; Sayols-Baixeras, S.; Aslibekyan, S.; Tiwari, H.K.; Irvin, M.R.; Bender, C.; Fei, D.; et al. Carbohydrate and fat intake associated with risk of metabolic diseases through epigenetics of CPT1A. *Am. J. Clin. Nutr.* **2020**, *112*, 1200–1211. [CrossRef]
57. Doaei, S.; Kalantari, N.; Izadi, P.; Salonurmi, T.; Jarrahi, A.M.; Rafieifar, S.; Azizi Tabesh, G.; Rahimzadeh, G.; Gholamalizadeh, M.; Goodarzi, M.O. Interactions between macro-nutrients' intake, FTO and IRX3 gene expression, and FTO genotype in obese and overweight male adolescents. *Adipocyte* **2019**, *8*, 386–391. [CrossRef]
58. Samblas, M.; Milagro, F.I.; Gomez-Abellan, P.; Martinez, J.A.; Garaulet, M. Methylation on the Circadian Gene BMAL1 Is Associated with the Effects of a Weight Loss Intervention on Serum Lipid Levels. *J. Biol. Rhythm.* **2016**, *31*, 308–317. [CrossRef]

59. Sehgal, R.; Perfilyev, A.; Mannisto, V.; Agren, J.; Nilsson, E.; Kakela, P.; Ling, C.; de Mello, V.D.; Pihlajamaki, J. Liver saturated fat content associates with hepatic DNA methylation in obese individuals. *Clin. Epigenetics* **2023**, *15*, 21. [CrossRef]
60. Bozack, A.K.; Rifas-Shiman, S.L.; Coull, B.A.; Baccarelli, A.A.; Wright, R.O.; Amarasiriwardena, C.; Gold, D.R.; Oken, E.; Hivert, M.F.; Cardenas, A. Prenatal metal exposure, cord blood DNA methylation and persistence in childhood: An epigenome-wide association study of 12 metals. *Clin. Epigenetics* **2021**, *13*, 208. [CrossRef]
61. Lindner, S.; Lucchini, R.; Broberg, K. Genetics and Epigenetics of Manganese Toxicity. *Curr. Environ. Health Rep.* **2022**, *9*, 697–713. [CrossRef]
62. Karnes, J.H.; Arora, A.; Feng, J.; Steiner, H.E.; Sulieman, L.; Boerwinkle, E.; Clark, C.; Cicek, M.; Cohn, E.; Gebo, K.; et al. Racial, ethnic, and gender differences in obesity and body fat distribution: An All of Us Research Program demonstration project. *PLoS ONE* **2021**, *16*, e0255583. [CrossRef] [PubMed]
63. McDaniel, T.; Wilson, D.K.; Coulon, M.S.; Sweeney, A.M.; Van Horn, M.L. Interaction of Neighborhood and Genetic Risk on Waist Circumference in African-American Adults: A Longitudinal Study. *Ann. Behav. Med.* **2021**, *55*, 708–719. [CrossRef] [PubMed]
64. Geetha, T. Novel Differentially Methylated Regions Identified by Genome-Wide DNA Methylation Analyses Contribute to Racial Disparities in Childhood Obesity. *Genes* **2023**, *14*, 1098. [CrossRef]
65. Dinour, L.M.; Bergen, D.; Yeh, M.C. The food insecurity-obesity paradox: A review of the literature and the role food stamps may play. *J. Am. Diet. Assoc.* **2007**, *107*, 1952–1961. [CrossRef]
66. Crawford, P.B.; Webb, K.L. Unraveling the paradox of concurrent food insecurity and obesity. *Am. J. Prev. Med.* **2011**, *40*, 274–275. [CrossRef]
67. Larson, N.I.; Story, M.T. Food insecurity and weight status among U.S. children and families: A review of the literature. *Am. J. Prev. Med.* **2011**, *40*, 166–173. [CrossRef]
68. Leung, C.W.; Epel, E.S.; Ritchie, L.D.; Crawford, P.B.; Laraia, B.A. Food insecurity is inversely associated with diet quality of lower-income adults. *J. Acad. Nutr. Diet.* **2014**, *114*, 1943–1953.e2. [CrossRef]
69. Leung, C.W.; Williams, D.R.; Villamor, E. Very low food security predicts obesity predominantly in California Hispanic men and women. *Public Health Nutr.* **2012**, *15*, 2228–2236. [CrossRef]
70. Vedovato, G.M.; Surkan, P.J.; Jones-Smith, J.; Steeves, E.A.; Han, E.; Trude, A.C.; Kharmats, A.Y.; Gittelsohn, J. Food insecurity, overweight and obesity among low-income African-American families in Baltimore City: Associations with food-related perceptions. *Public Health Nutr.* **2016**, *19*, 1405–1416. [CrossRef]

nutrients

MDPI

Article

Height-Related Polygenic Variants Are Associated with Metabolic Syndrome Risk and Interact with Energy Intake and a Rice-Main Diet to Influence Height in KoGES

Sunmin Park

Department of Food and Nutrition, Obesity/Diabetes Research Center, Hoseo University, 165 Sechul-Ri, BaeBang-Yup, Asan-Si 336-795, ChungNam-Do, Republic of Korea; smpark@hoseo.edu; Tel.: +82-41-540-5633; Fax: +82-41-540-5638

Abstract: Adult height is inversely related to metabolic syndrome (MetS) risk, but its genetic impacts have not been revealed. The present study aimed to examine the hypothesis that adult height-related genetic variants interact with lifestyle to influence adult height and are associated with MetS risk in adults aged >40 in Korea during 2010–2014. Participants were divided into short stature (SS; control) and tall stature (TS; case) by the 85th percentile of adult height. The genetic variants linked to adult height were screened from a genome-wide association study in a city hospital-based cohort (n = 58,701) and confirmed in Ansan/Ansung plus rural cohorts (n = 13,783) among the Korean Genome and Epidemiology Study. Genetic variants that interacted with each other were identified using the generalized multifactor dimensionality reduction (GMDR) analysis. The interaction between the polygenic risk score (PRS) of the selected genetic variants and lifestyles was examined. Adult height was inversely associated with MetS, cardiovascular diseases, and liver function. The PRS, including zinc finger and BTB domain containing 38 (*ZBTB38*)_rs6762722, polyadenylate-binding protein-interacting protein-2B (*PAIP2B*)_rs13034890, carboxypeptidase Z (*CPZ*)_rs3756173, and latent-transforming growth factor beta-binding protein-1 (*LTBP1*)_rs4630744, was positively associated with height by 1.29 times and inversely with MetS by 0.894 times after adjusting for covariates. In expression quantitative trait loci, the gene expression of growth/differentiation factor-5 (*GDF5*)_rs224331, non-SMC condensin I complex subunit G (*NCAPG*)_rs2074974, ligand-dependent nuclear receptor corepressor like (*LCORL*)_rs7700107, and insulin-like growth factor-1 receptor (*IGF1R*)_rs2871865 was inversely linked to their risk allele in the tibial nerve and brain. The gene expression of *PAIP2B*_rs13034890 and a disintegrin and metalloproteinase with thrombospondin motifs-like-3 (*ADAMTSL3*)_rs13034890 was positively related to it. The PRS was inversely associated with MetS, hyperglycemia, HbA1c, and white blood cell counts. The wild type of *GDF5*_rs224331 (Ala276) lowered binding energy with rugosin A, D, and E (one of the hydrolyzable tannins) but not the mutated one (276Ser) in the in-silico analysis. The PRS interacted with energy intake and rice-main diet; PRS impact was higher in the high energy intake and the low rice-main diet. In conclusion, the PRS for adult height interacted with energy intake and diet patterns to modulate height and was linked to height and MetS by modulating their expression in the tibial nerve and brain.

Keywords: stature; *GDF5*; *IGF-1R*; polygenic risk score; in silico analysis; energy intake

1. Introduction

Adult height is achieved by growth during childhood and adolescence, with full height attained at age 18–20. The adult height of a child or adolescent between the age of 4 to 17.5 years can be predicted based on the current height, body mass, chronological age, and parents' height. The genetic impact on height increases can be estimated from adult height. However, after 30 years of age, people, especially women, gradually lose height. A loss of height of about 2.54 cm for men and 5.08 cm for women occurs between ages 30 to 70 years. However, a remarkable decline in height indicates certain health problems. Loss of height with age may occur due to poor nutrition, compression, and dehydration of the discs between the vertebrae, curvature of the spine, low bone mineral density, diabetes, and low muscle mass in the torso [1]. Therefore, when adjusted for osteoporosis, skeletal muscle mass, nutritional status, and age, the current adult height would reflect height in the 20s [2]. The genetic variants associated with adult height can be analyzed through genome-wide association studies (GWAS) after adjusting for age, gender, osteoporosis, income, body mass index (BMI), education, and energy intake. The interaction between genetics and lifestyle factors can be evaluated.

Human height is modulated by genetic factors that determine about 80 percent of an individual's height, and it is a polygenic trait. The GWAS of adult height has shown that common variants account for 50% of height variations [3,4]. However, the influencing genetic variants are only partially understood. Studies determining genetic variants associated with height have been conducted in not only children but also adults. The genetic variants related to adult height were identified as solute carrier family-27 (*SLC27A3*) and cytochrome P450, family 26, subfamily B (*CYP26B1*) in Japan [5], and insulin-like growth factor (*IGF)-2/H19* and Tet methylcytosine dioxygenase-1 (*TET1*) in Icelanders [6]. In a study on 8842 Koreans, several adult height-related loci were identified, such as exostosin glycosyltransferase-1 (*EXT1*), FRAS1-related extracellular matrix-1 (*FREM1*), paralemmin-2 (*PALM2*), A-kinase anchor protein-2 (*AKAP2*), nucleoporin-37 (*NUP37*), pro-melanin concentrating hormone (*PMCH*), *IGF1*, keratin-20 (*KRT20*), and ankyrin repeat domain-60 (*ANKRD60*). Their risk alleles explain about 1.0% of the height variations among Koreans [7]. In a pediatric cohort with European ancestry, 16 genetic variants in epidermal growth factor containing fibulin extracellular matrix protein-1 (*EFEMP1*)-polyribonucleotide nucleotidyltransferase-1 (*PNPT1*), G protein-coupled receptor-126 (*GPR126*), *C6orf173*, sperm associated antigen-17 (*SPAG17*), *histone class-1*, human leukocyte antigen (*HLA) class III*, and growth differentiation factor-5 (*GDF5*)-ubiquinol-cytochrome c reductase complex chaperone CBP3 homolog (*UQCC*) were associated with height, which explains 1.64% of the variations in height [8]. In Chinese children, zinc finger and BTB domain containing 38 (*ZBTB38*), zinc finger protein 638 (*ZNF638*), ligand-dependent nuclear receptor corepressor like (*LCORL*), cyclin-dependent kinase10 (*CDK10*), *Cdk5 and Abl enzyme substrate 1* (*CABLES1*), and tRNA-splicing endonuclease subunit-15 (*TSEN15*) related genetic variants were revealed to be linked to height [9].

Adult height is demonstrated to be associated with several metabolic and other diseases. In one study, adult height is found to have an inverse relationship with 2-h serum glucose concentrations only in adults with <35 kg/m^2 BMI during oral glucose tolerance tests [10], suggesting the height effects can be offset with adiposity to influence blood glucose levels. Furthermore, the genetic variants related to height have shown positive and inverse associations with various metabolic diseases, such as atrial fibrillation (OR = 1.33, 95% CI = 1.26–1.40) [11], venous thromboembolism (OR = 1.15, 95% CI = 1.11–1.19), hip fracture (OR = 1.27, 95% CI = 1.17–1.39), hypertension (OR = 0.88, 95% CI = 0.85–0.91), and coronary artery disease (OR = 0.86, 95% CI = 0.82–0.90) [12]. Multiple height-associated pathways link height with various diseases indicating that multiple biological mechanisms affecting height raise the risk of these diseases [12]. Therefore, height and metabolism may be linked to height-related genetic variants.

Few studies have examined the association between height-related genetic variants and metabolic syndrome (MetS). It was hypothesized that the genetic variants related to adult height were associated with MetS risk and that they interacted with lifestyle factors. The purpose of the current study was to examine the hypothesis by (1) exploring the genetic variants linked to tall stature after adjusting for the factors associated with a reduction in height among adults, (2) generating the optimal genetic model for tall stature, (3) analyzing the association of polygenic risk scores (PRS) with the risk of MetS and its components, and (4) examining the interaction of PRS with lifestyle factors to affect adult height in the city hospital-based cohort.

2. Methods and Materials

2.1. Participants

Adult volunteers aged over 40 years were recruited in a city-hospital cohort (n = 58,701), the Ansan/Ansung cohort (n = 5493), and a rural cohort (n = 8105) in the Korean Genome and Epidemiology Study (KoGES) during the period 2010–2014 [13]. The objectives were primarily evaluated in the city-hospital cohort in the present study, and the Ansan/Ansung plus rural cohorts were applied as replicate studies for GWAS results for height. The institutional review board (IRB) of the Korea National Institute of Health approved the KoGES (KBP-2015-055), and the IRB of Hoseo University approved the present study (HR-034-01). All participants signed a written informed consent form.

2.2. Adult Height Criteria

The short stature (SS; n = 51,165; control) and tall stature (TS; n = 7536; case) groups of the participants were divided by the cutoff of ≥175 cm for men and ≥163 cm for women as tall stature, and below that for short stature. These heights were the 85th percentiles of adult height for people aged >40 between 2010–2014 in Korea.

2.3. Survey Questionnaires and Anthropometric and Biochemical Measurements

On their first visit, the participants completed the demographic questionnaires. Height and weight were measured with an automatic digital machine wearing light gowns and bare feet. The appendicular skeletal muscle mass in the city hospital cohort was predicted using a prediction model generated by the XGBoost algorithm in the Ansan/Ansung cohort [14]. Blood pressure was determined with a sphygmomanometer in a sitting position three times, and average systolic (SBP) and diastolic blood pressure (DBP) were recorded. The participants fasted for over 12 h. The serum and plasma were collected from the blood drawn without and with ethylenediaminetetraacetic acid (EDTA) and heparin, respectively. Biochemical parameters in the serum were measured with previously described standard methods [15]. They included glucose, triglyceride, total cholesterol, HDL, alanine aminotransferase (ALT), aspartate aminotransferase (AST), and creatinine concentrations in the serum, and blood hemoglobin A1c (HbA1c) and white blood cell (WBC) contents. The estimated glomerular filtration rate (eGFR) was calculated using the equation generated by the Modification of Diet in Renal Disease study [16]. The insulin resistance was calculated using the prediction model generated by the homeostatic model assessment of insulin resistance (HOMA-IR) in the Ansan/Ansung cohort from a previous study [17].

Details of lifestyle factors, including food intake, were collected using questionnaires during a health interview by a trained technician. Daily alcohol and coffee intakes were calculated by multiplying their intake amounts with frequencies. Smoking status was classified into non-, past, and current smokers. Smokers were specified as smoking >100 cigarettes in their lifetime, and current smokers who had not smoked for 6 months were considered past smokers. Exercise status was classified into regular exercise and no exercise based on the cutoff of 30 min of moderate-intensity activity such as water pushing a lawn mower, brisk walking, hiking, aerobics, riding a bike, dancing, doubles tennis, or rollerblading for ≥3 days per week.

2.4. Usual Food Intake Measurement

The usual food and nutrient consumption during the last 12 months was determined based on the response to a semi-quantitative food frequency questionnaire (SQFFQ) generated for Koreans. It included 106 foods Koreans commonly consumed. The SQFFQ was confirmed with three-day food records for the four seasons in a previous study [17]. The food frequencies were divided into 8 categories, such as seldom, once per month, two to three times per month, once or twice per week, three or four times per week, five or six times per week, daily, twice a day, and ≥3 times a day. One portion size of each food was classified into either more than, equal to, or half of the portion size shown in a photograph. Participants selected the frequencies and one portion size of each food item. Their intake was calculated by multiplying the median of the frequencies by one portion size of each food. Daily nutrient intake was estimated from the calculated daily food amount using the computer-aided nutritional analysis program CAN-Pro 2.0 (Korean Nutrition Society, Seoul, Republic of Korea). Daily nutrient intake was calculated by summing the individual nutrient of each food.

2.5. Dietary Patterns and Dietary Inflammatory Index (DII)

The dietary patterns were clustered with principal component analysis (PCA) of 30 predefined groups categorized from the 106 food items. Four dietary patterns were designated based on eigenvalues >1.5, orthogonal rotation procedure (varimax), and ≥0.40 factor-loading values. The four dietary patterns were named as Korean-balanced diet (KBD), Western-style diet (WSD), plant-based diet (PBD), and rice-main diet (RMD), according to foods. The dietary inflammation index (DII) was calculated with a prediction equation generated with the food and nutrient intake and divided by 100 [18].

2.6. Genotyping, Its Quality Control, and Genotype-Tissue Expression (GTEx)

Genomic DNA was isolated from the volunteers' blood and genotyping was performed using a Korean Chip (Affymetrix, Santa Clara, CA, USA). It was made to evaluate the genetic impact of single nucleotide polymorphisms (SNPs) on metabolic diseases in Koreans at the Center for Genome Science at the Korea National Institute of Health [19]. The genotypes were imputed based on the Korean Haplotype Map (HapMap) data [20]. The genotyping accuracy was assessed with a Bayesian learning algorithm for Robust General Linear Models (RGLMs) [21]. Genotyping accuracy was checked with the exclusion criteria as follows: <98% genotyping accuracy, ≥30% heterozygosity, ≥1% missing genotype call rate, ≤1% minor allele frequency (MAF), and $p \leq 0.05$ Hardy–Weinberg equilibrium (HWE) [21]. The significance distribution of the genetic variants was visualized with a Manhattan plot by the R program. A Q–Q probability plot indicates the goodness of fit of the actual data distribution to the theoretical data distribution. When the lambda value in the Q–Q plot was close to 1, the genotypes were ideal. Among the selected genotypes, the pathway associated with height was selected with satisfying the P value for Bonferroni correction using the multi-marker analysis of genomic annotation (MAGMA) gene-set analysis in the SNP2GENE function of the functional mapping and annotation (FUMA) web application (https://github.com/Kyoko-wtnb/FUMA-webapp/, accessed on 20 April 2022).

Genotype-Tissue Expression (GTEx) was used to show the gene expression with different genetic variant alleles in various tissues using the GTEx expression quantitative trait loci (eQTL) calculator (https://gtexportal.org/home/testyourown, accessed on 3 May 2022).

2.7. Selection of Genetic Variants to Influence Adult Height and Their Optimal Model with the SNP–SNP Interaction

Figure 1 provides the optimal genetic model selection process. First, genetic variants (n = 1299) influencing height were selected using the SS and TS groups in the city hospital-based cohort at a significance level of $p < 5 \times 10^{-7}$, as defined above. Of these, 956 genetic variants satisfied MAF (>1%) and HWE ($p > 0.05$), and the 79 unduplicated gene names were identified using g:Profiler (https://biit.cs.ut.ee/gprofiler/snpense, accessed on 7 March 2022). The genetic variants with high linkage disequilibrium (LD; D′ ≥ 0.2) were excluded due to giving the same genetic impact on the height. After exclusions, 37 genetic variants remained to meet the criteria (D′ < 0.2) using Haploview 4.2 in the PLINK toolset, and 15 with the same or unidentified gene names were eliminated.

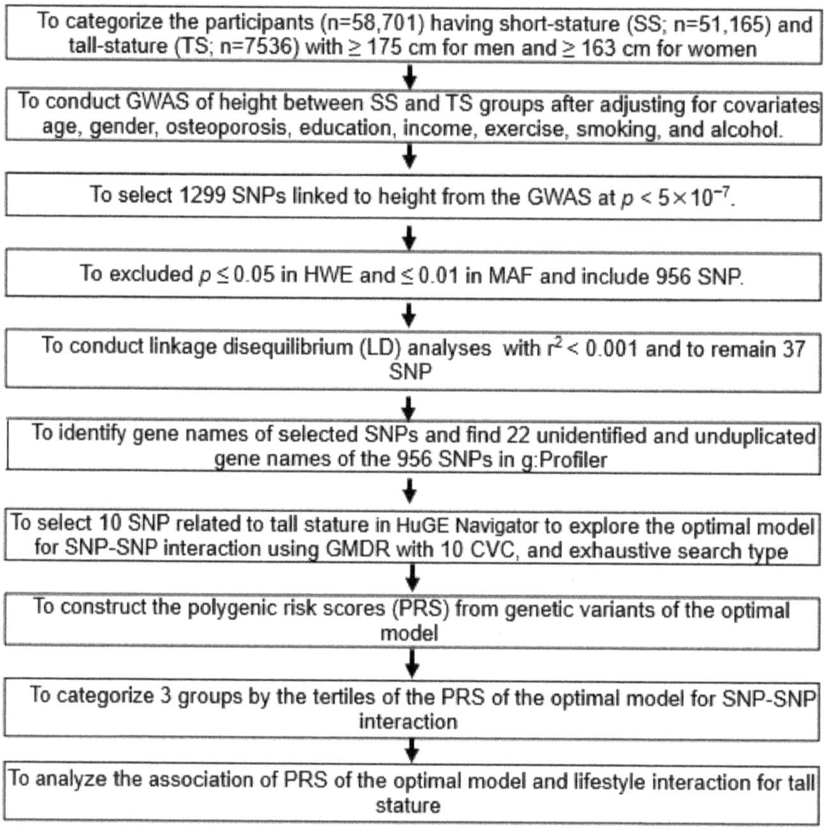

Figure 1. Flow chart to generate polygenic risk score (PRS) associated with stature by SNP–SNP interaction and its interaction with lifestyle factors. The short stature (SS; n = 51,165; control) and tall stature (TS; n = 7536; case) groups of the participants were divided by the cutoff of ≥175 cm for men and ≥163 cm for women, which were the 15th percentiles of adult height in people aged >40 years in 2010–2014 in Korea.

Ten SNPs were included in the optimal genetic model by selecting them based on their interaction with each other from 22 genetic variants in the generalized multifactor dimensionality reduction (GMDR) software version 2.0 using the exhaustive search type after adjusting for covariates. Covariate set 1 contained age, gender, residence area, education, and income, and covariate set 2 included covariates in set 1 plus energy intake, alcohol intake, regular exercise, and smoking status for model 2. The selection criteria were a $p < 0.001$ for the sign test of the testing balanced accuracy (TEBA) and ten cross-validation consistency (CVC) in the ten-fold cross-validation [22]. Among the optimal models to satisfy the significance level of the sign test and CVC, the model with the lowest number of genetic variants and highest odds ratios (ORs) was selected as the optimal model.

The PRS for the optimal genetic model was calculated as a sum of the number of the risk alleles from each selected SNP in the optimal SNP–SNP interaction model [23–25]. As the SNP risk allele was G, the genetic scores of "AA", "AG", and "GG" were 0, 1, and 2, respectively. Each model was divided into Low-, Middle-, and High-PRS. The PRS in the four-SNP model was classified into Low-, Middle-, and High-PRS based on scores of 0–3, 4–5, and ≥6, respectively; and that in the seven-SNP model was categorized into 2–7, 8–10, and ≥11, respectively. The PRS of the model was further used for analyzing the interaction with the lifestyle parameters.

2.8. Molecular Docking of Wild and Mutated GDF5 with Food Compounds

The wild and mutated chemical structures of the proteins *GDF5_rs224331* (Ala276Ser) were made in the Protein Data Bank (PDB) format from the Iterative Threading Assembly Refinement (I-TASSER) website (https://zhanggroup.org/I-TASSER/ (accessed on 20 June 2022). Food compounds (n = 20,000) were downloaded from the fooDB website (https://foodb.ca/, accessed on 9 June 2022). Water molecules attached to the ligands were eliminated using the pleomorphic analysis methodology (PyMOL) software version 2.0 (DeLano Scientific LLC, South San Francisco, CA, USA) [26]. The PDB structure of the protein and ligand was changed into a protein data bank, partial charge (Q), and atom type (T) (PDBQT) files lattice format using AutoDock Tools 1.5.6 (Molecular Graphics Laboratory, Scripps Research Institute, La Jolla, CA, USA) [26]. The active sites of the proteins GDF5 were identified using the ProteinsPlus website (https://proteins.plus/, accessed on 8 July 2022), and they and the mutated sites (rs224331) were included for molecular docking.

Food components with a <−10 kcal/mol binding free energy were selected [27]. The binding free energy at the active site indicated the binding affinity of *GDF5* to the food compound. The lower the binding free energy, the tighter the binding and affinity.

2.9. Molecular Dynamics Simulation (MDS)

The conformational changes in the GDF5 structures were examined using MDS to detect the changes in their activity. After the top docking poses with the selected food compounds were inserted, simulations were conducted on the *GDF5* active site and docked complexes. Additionally, the Chemistry at Harvard Macromolecular Mechanics (CHARMM) force field was added to each molecular structure generated by "Simulation", and the protein was solvated by "Solvation". The "Standard Dynamics Cascade" was used to set the molecular dynamics simulation parameters for the protein added to the solvent system. The ramp-up time, equilibration time, simulation sampling time, and simulation step size were set to 40 ps, 400 ps, 10,000 ps, and 2 fs, respectively, and other parameters were set to default values. The hydrogen bond values, root mean square deviation (RMSD), and root mean square fluctuations (RMSF) were analyzed after the 10 ns simulation.

2.10. Statistical Analysis

The sample size (n = 58,701) was sufficient to show significance at $\alpha = 0.05$ and $\beta = 0.99$ when assigned an odds ratio (OR) of 1.08 in the logistic regression analysis using a G-power calculator. Descriptive statistical analysis was conducted using SAS (version 9.3; SAS Institute, Cary, NC, USA). Frequency distributions were used for the categorical variables, and their statistical differences between the SS and TS groups were calculated with the Chi-square test. The adjusted means and standard deviations for the continuous variables were assessed after adjusting the covariates, including age, gender, education, income, survey year, BMI, osteoporosis, energy intake, exercise, alcohol intake, and smoking status. The statistical differences between the genders and adult height groups were analyzed using a two-way analysis of covariance (ANCOVA) with adjustment for the covariates [28]. When ANCOVA was significant, multiple comparisons according to the genders and height groups were performed using Tukey's test.

The association of height with anthropometric and biochemical parameters was analyzed using logistic regression analysis after adjustment for covariates. The results are provided with each metabolic parameter's adjusted odds ratios (ORs) and 95% confidence intervals (CI). Two different covariate sets were included according to the covariates. Covariate set 1 included age, place of residence, survey year, osteoporosis, BMI, education, and income as covariates. Covariate set 2 was calculated with covariates of model 1 plus energy intake, physical activity, smoking status, and alcohol intake.

The two-way ANCOVA was conducted with the main effect terms of PRS and lifestyles, their interaction term, and covariates. If the interaction terms showed statistical significance, each lifestyle parameter was categorized into high or low groups according to the dietary reference intake [29] or their 30th percentile. The ORs and 95% CI of the height with PRS were also assessed with logistic regression analysis in each of the high or low groups in the lifestyle-related parameters. Any significant difference in the height was determined according to the PRS groups using the χ^2 test or ANCOVA in the lifestyle-related parameters' low- or high-groups.

3. Results

3.1. Demographic Characteristics and Lifestyles According to Gender and Adult Height

The age was higher in the SS group than in the TS group in both genders (Table 1). The proportion of tall persons was higher than that of short persons in men, and it was the opposite in women. The proportion of both men and women in the TS who lacked higher education (≤middle school) was much lower than in the SS group (Table 1). However, the proportion of men with a high income was lower in the TS than in the SS group, but it was the reverse in women. Smoking status was not significantly different between the SS and TS groups in both genders (Table 1). The proportion of the participants with regular physical exercise was small but significantly higher in the TS group than in the SS group, but only in women. Only in men, alcohol intake was higher in the TS group than in the SS group (Table 1).

Table 1. Demographic characteristics of the participants according to gender and adult height.

	Men		Women	
	Short-Stature (n = 17,305)	**Tall-Stature (n = 2988)**	**Short-Stature (n = 33,860)**	**Tall-Stature (n = 4548)**
Age (years)	56.4 ± 0.06^{b}	53.2 ± 0.18^{a}	52.8 ± 0.04^{c}	$50.2 \pm 0.11^{c***+++}$
Gender (%)	33.82	39.65 ‡‡‡	66.18	60.35 ‡‡‡
Education				
≤Middle school	1654 (15.1)	99 (6.52) ‡‡‡	6409 (23.5)	329 (10.5) ‡‡‡
High school	8261 (75.4)	1173 (77.3)	19,373 (71.1)	2498 (80.0)
≥College	1045 (9.53)	246 (16.2)	1486 (5.45)	296 (9.48)
Income				
≤$2000	1483 (9.03)	268 (10.7) ‡‡‡	3911 (12.3)	261 (6.02) ‡‡‡
$2000–4000	7184 (43.7)	1125 (44.9)	14,278 (44.9)	1707 (39.3)
>$4000	7764 (47.3)	1111 (44.4)	13,599 (42.8)	2371(54.6)
Former smoking (%)	2237 (12.9)	349 (11.7)	100 (0.30)	17 (0.37)
Smoking (%)	1447 (8.36)	275 (9.21)	172 (0.51)	19 (0.42)
Physical exercise (%)	10,168 (59.0)	1784 (59.9)	17,579 (52.1)	2445 (53.8) ‡
Alcohol (g)	35.2 ± 0.38^{b}	39.8 ± 0.88^{a}	5.24 ± 0.27^{b}	$5.89 \pm 0.71^{b***+++\#\#}$
Energy intake (EER %)	89.4 ± 0.26^{d}	93.3 ± 0.58^{c}	99.1 ± 0.18^{b}	$101.6 \pm 0.47^{a***\#\#\#}$
Carbohydrates (En%)	71.3 ± 0.06^{b}	71.2 ± 0.13^{b}	72.0 ± 0.04^{a}	$71.6 \pm 0.11^{b***+}$
Fat (En%)	14.2 ± 0.04^{a}	14.3 ± 0.10^{a}	13.7 ± 0.03^{b}	$14.1 \pm 0.08^{a***++\#}$
Protein (En%)	13.3 ± 0.02^{b}	13.3 ± 0.05^{b}	13.5 ± 0.02^{a}	$13.5 \pm 0.04^{a***}$
Fiber (g)	15.2 ± 0.08^{b}	15.8 ± 0.18^{a}	14.3 ± 0.06^{c}	$14.6 \pm 0.15^{c***+++}$
Calcium (mg)	411 ± 2.15^{d}	427 ± 4.89^{c}	456 ± 1.50^{b}	$471 \pm 3.95^{a***+++}$
Vitamin C (mg)	95.1 ± 0.56^{d}	98.9 ± 1.28^{c}	110 ± 0.39^{b}	$113 \pm 1.04^{a***++}$
Vitamin D (ug)	5.59 ± 0.05^{d}	5.89 ± 0.11^{c}	6.81 ± 0.03^{b}	$7.09 \pm 0.09^{a***+++}$
DII (scores)	-13.8 ± 0.39	-14.2 ± 0.89	-13.4 ± 0.27	-15.5 ± 0.72
Flavonoids (mg)	32.1 ± 0.27^{b}	32.8 ± 0.61^{b}	41.5 ± 0.19^{a}	$42.6 \pm 0.49^{a***+}$
KBD (%)	6832 (39.5)	1269 (42.5) ‡‡	10,059 (29.7)	1409 (31.0)
PBD (%)	3551 (20.5)	647 (21.7)	13,316 (39.3)	2062 (45.3) ‡‡‡
WSD (%)	8619 (49.8)	1812 (60.6) ‡‡‡	11,155 (32.9)	1966 (43.2) ‡‡‡

Table 1. *Cont.*

	Men		Women	
	Short-Stature (n = 17,305)	**Tall-Stature (n = 2988)**	**Short-Stature (n = 33,860)**	**Tall-Stature (n = 4548)**
RMD (%)	5482 (31.7)	983 (32.9)	11,480 (33.9)	1626 (35.8) ‡
Coffee intake (g/day)	4.23 ± 0.03 [a]	4.25 ± 0.06 [a]	3.34 ± 0.02 [c]	3.48 ± 0.05 [b***+]
Tea (g/day)	43.5 ± 0.71 [ab]	47.7 ± 1.63 [a]	43.2 ± 0.50 [b]	42.2 ± 1.32 [b***]

Values represent adjusted means and standard errors for continuous variables and the number and percentage for categorical variables. Values represent adjusted odd ratios and 95% confidence intervals. Covariates included age, sex, weight at age 18, education, income, energy intake (percentage of estimated energy requirement), residence areas, daily activity, alcohol intake, and smoking status. EER, estimated energy requirement; En%, energy percentage; DII, dietary inflammation index; KBD, Korean-balanced diet; PBD, plant-based diet; WSD, Western-style diet; RMD, rice-main diet. *** Significant differences by gender at $p < 0.001$. + Significant differences by height at $p < 0.05$, ++ at $p < 0.01$, +++ $p < 0.001$. # Significant interaction between gender and height at $p < 0.05$, ## at $p < 0.01$, ### $p < 0.001$. ‡ Significantly different from the control group in χ^2 test in each gender at $p < 0.05$, ‡‡ at $p < 0.01$, ‡‡‡ at $p < 0.001$. [a, b, c, d] Different superscript letters indicated significant differences among the groups in Tukey's test at $p < 0.05$.

3.2. Nutrient Intake According to Gender and Adult Height

Energy intake was lower in the SS than in the TS group in both genders. Regarding the intake of macronutrients, fat intake was lower in the SS group than in the TS group only in women, and the trend of carbohydrate intake was the opposite of that of the fat intake (Table 1). However, protein intake was not significantly different between the SS and TS groups for both genders. Dietary fiber intake was lower in the SS group than in the TS group only in men, and calcium and vitamin C and D intakes were lower in the SS group than in the TS group for both genders (Table 1). Intakes of DII and flavonoids did not differ between the TS and SS groups (Table 1). The proportion of KBD was lower in the SS group than in the TS group only in men, but that of PBD and RMD was lower in the SS group than in the TS group only in women (Table 1). The proportion of WSD was much lower in the SS group than in the TS group in both genders. Coffee intake, but not tea intake, was lower in the SS group than in the TS group in women (Table 1).

3.3. Prevalence of MetS and Its Related Parameters According to Gender and Adult Height

As expected, height was much greater in the TS group than in the SS group in both genders. When the increase in height ceased at age 18, the weight was not significantly different between the TS and SS groups (Table 2). In women, the BMI during the study was higher in the SS group than the TS group and was inversely associated with height (adjusted ORs = 0.908). Waist circumferences were lower in the TS group than in the SS group for both genders. The inverse association of waist circumference with height was much more significant than the association with BMI (adjusted ORs = 0.327) (Table 2). The skeletal muscle index (SMI) and fat mass were significantly lower in the TS group than the SS group in both genders and was inversely associated with height. White blood corpuscles (WBC), which reflect the immune status, showed a trend similar to the SMI and fat mass (Table 2). However, there was no significant difference in serum high-sensitive C-reactive protein (hs-CRP) concentrations, an inflammatory marker, between the TS and SS groups (Table 2).

MetS and cardiovascular disease (CVD) incidence was inversely associated with height. Fasting serum glucose concentrations and blood HbA1c levels were higher in the SS group than in the TS group for both genders, and they showed an inverse relationship with height (Table 2). Insulin resistance also showed an inverse association with height (Table 2). Dyslipidemia was also inversely related to height, showing that short persons had hyper-total cholesterol, hypo-high-density lipoprotein (HDL), hyper-low-density lipoprotein (LDL), and hyper-triglycerides in the blood, compared to tall persons (Table 2). SBP and DBP were higher in the SS group than in the TS group and revealed an inverse association with height (Table 2). Serum alanine aminotransferase (ALT) and aspartate aminotransferase (AST), the indexes of liver damage, had an inverse association with height. However, the eGFR estimated with serum creatinine concentrations was not significantly linked to height (Table 2). Furthermore, arthritis incidence, but not osteoporosis, was inversely associated with height (Table 2).

Table 2. Adjusted means and odds ratio (ORs) of metabolic syndrome-related parameters according to gender and adult height.

	Men		Women		
	Short-Stature (n = 17,305)	**Tall-Stature (n = 2988)**	**Short-Stature (n = 33,860)**	**Tall-Stature (n = 4548)**	**Adjusted OR and 95% CI**
Height (cm) [1]	167.3 ± 0.03^{b}	177.2 ± 0.08^{a}	155.5 ± 0.02^{d}	$164.7 \pm 0.06^{c***+++\#\#\#}$	
BMI (kg/m²) [2]	24.5 ± 0.04^{a}	24.6 ± 0.07^{a}	23.6 ± 0.03^{b}	$23.0 \pm 0.06^{c***+++\#\#\#}$	0.908 (0.857–0.962)
Waist (cm) [3]	80.1 ± 0.07^{c}	76.9 ± 0.13^{a}	81.5 ± 0.05^{b}	$78.7 \pm 0.10^{d***+++\#}$	0.327 (0.299–0.357)
Weight at age 18 (kg) [4]	55.4 ± 0.14	56.1 ± 0.31	55.0 ± 0.11	$55.1 \pm 0.20^{***}$	1.025 (0.956–1.099)
SMI (kg/m) [5]	7.34 ± 0.004^{a}	7.04 ± 0.010^{b}	6.90 ± 0.003^{c}	$6.46 \pm 0.006^{d***+++\#}$	0.402 (0.376–0.430)
Fat mass (%) [6]	20.4 ± 0.02^{c}	17.3 ± 0.05^{d}	32.8 ± 0.02^{a}	$30.1 \pm 0.04^{b***+++\#\#\#}$	0.479 (0.445–0.515)
WBC (10^9/L) [7]	5.79 ± 0.02^{a}	5.61 ± 0.04^{b}	5.67 ± 0.01^{b}	$5.60 \pm 0.03^{c*+++\#}$	0.813 (0.767–0.861)
hs-CRP (mg/dL) [8]	0.14 ± 0.004^{ab}	0.16 ± 0.009^{a}	0.14 ± 0.003^{ab}	$0.12 \pm 0.007^{b***}$	0.784 (0.612–1.005)
MetS [9]	3005 (17.4)	593 (19.9) ‡‡	4275 (12.6)	427 (9.39) ‡‡‡	0.494 (0.452–0.540)
CVD [9]	1107 (6.41) ‡‡	109 (3.65)	1030 (3.05)	68 (1.50) ‡‡‡	0.669 (0.563–0.794)
Glucose (mg/dL) [10]	96.38 ± 0.27^{a}	93.4 ± 0.55^{b}	95.4 ± 0.19^{a}	$93.4 \pm 0.42^{b+++}$	0.718 (0.657–0.785)
HbA1c (%) [11]	5.63 ± 0.01^{b}	5.49 ± 0.03^{c}	5.79 ± 0.01^{a}	$5.68 \pm 0.02^{b***+++}$	0.659 (0.569–0.763)
Insulin resistance (%) [9]	1955 (11.3)	347 (12.0)	20,66 (6.1)	226 (4.97) ‡‡	0.542 (0.487–0.603)
Total cholesterol (mg/dL) [12]	189 ± 0.35^{c}	190 ± 0.71^{c}	202 ± 0.23^{a}	$198 \pm 0.56^{b***+++\#\#\#}$	0.707 (0.658–0.758)
HDL (mg/dL) [13]	52.2 ± 0.18^{b}	53.9 ± 0.36^{c}	55.1 ± 0.13^{a}	$57.3 \pm 0.24^{a***++\#\#\#}$	1.330 (1.249–1.415)
LDL (mg/dL) [14]	112 ± 0.45^{c}	112 ± 0.94^{c}	122 ± 0.33^{a}	$117 \pm 0.71^{b***+++\#\#\#}$	0.702 (0.646–0.763)
TG (mg/dL) [15]	120 ± 1.11^{b}	105 ± 2.29^{c}	127 ± 0.81^{a}	$109 \pm 1.74^{c**+++}$	0.633 (0.594–0.675)
SBP (mmHg) [16]	123 ± 0.20^{a}	120 ± 0.40^{b}	123 ± 0.14^{a}	$120 \pm 0.31^{b+++}$	0.749 (0.704–0.796)
DBP (mmHg) [17]	76.9 ± 0.13^{a}	75.1 ± 0.27^{b}	75.5 ± 0.09^{b}	$73.9 \pm 0.20^{c***++\#}$	0.698 (0.633–0.770)
AST (IU/L) [18]	25.1 ± 0.20^{a}	25.1 ± 0.45^{a}	23.1 ± 0.14^{b}	$22.5 \pm 0.36^{b***}$	0.633 (0.550–0.729)
ALT (IU/L) [19]	24.5 ± 0.16^{a}	23.6 ± 0.34^{b}	23.5 ± 0.12^{b}	$22.4 \pm 0.25^{c**+++}$	0.531 (0.485–0.582)
Egfr [20]	84.4 ± 0.21^{b}	84.1 ± 0.43^{b}	86.6 ± 0.15^{a}	$85.3 \pm 0.33^{b***++}$	0.940 (0.867–1.020)
Arthritis (N, %) [9]	698 (4.04)	111 (3.72)	3995 (11.8)	326 (7.17) ‡‡‡	0.868 (0.774–0.973)
Osteoporosis (N, Yes%) [9]	117 (0.68)	16 (0.54)	2779 (8.22)	162(3.56) ‡‡‡	0.882 (0.740–1.051)

CI, confidence interval; SBP, systolic blood pressure; DBP, diastolic blood pressure; AST, aspartate aminotransferase; ALT, alanine aminotransferase; TG, triglyceride. The high-height group was ≥175 for men and ≥163 for women. Adjusting for covariates including weight, gender, age, weight at age 18, education, income, total activity, energy intake, alcohol intake, residence area, and smoking status. The cutoff points of the reference for logistic regression were as following: [1] <172.5 cm for men and <160 cm for women; [2] <25 kg/m² for BMI; [3] <90 cm for men and 85 cm for women waist circumferences; [4] <60 kg for men and <59 kg for women; [5] <7.6 kg/m in men and 5.4 kg/m in women for SMI in skeletal muscle index (SMI defined as appendicular skeletal muscle mass/height); [6] <25% for men and 30% for women for fat mass; [7] <4.0 × 10^9 counts/L WBC; [8] <0.5 mg/dL serum high-sensitive C-reactive protein (hs-CRP) concentrations; [9] disease incidence; [10] <126 mL/dL fasting serum glucose plus diabetic drug intake; [11] <6.5% blood HbA1c plus diabetic drug intake; [12] <230 mg/dL serum total cholesterol concentrations; [13] >40 mg/dL for men and 50 mg/dL for women serum HDL cholesterol; [14] <160 mg/dL serum total cholesterol concentrations; [15] <150 mg/dL serum triglyceride concentrations; [16] <140 mmHg SBP, [17] <90 mmHg DBP plus hypertension medication; [18] <40 U/L aspartate aminotransferase; [19] <35 U/L alanine aminotransferase; [20] <70 mL/min/1.73 m² estimated glomerular filtration rate (eGFR). * Significant differences by gender at $p < 0.05$, ** at $p < 0.01$, *** at $p < 0.001$. ++ Significant differences by height at $p < 0.01$, +++ $p < 0.001$. # Significant interaction between gender and height at $p < 0.05$, ### at $p < 0.001$. ‡‡ Significantly different from the control group in χ^2 test in each gender at $p < 0.01$, ‡‡‡ at $p < 0.001$. a, b, c, d Different superscript letters indicated significant differences among the groups in Tukey's test at $p < 0.05$.

3.4. Genetic Variants Linked to Adult Height

The genetic variants affecting adult height are shown as a Manhattan plot in Supplementary Figure S1A, and those satisfying the selection criteria were located in chromosomes 2, 3, 5, 6, 12, 15, and 20. The Q–Q plot in Supplementary Figure S1B indicated that the genetic variants did not deviate from the expected values (lambda = 1.032).

The genetic variants that interacted with each other to promote an increase in height are presented in Table 3. Polyadenylate-binding protein-interacting protein 2B (*PAIP2B*)_rs13034 890, carboxypeptidase Z (CPZ)_rs3756173, non-SMC condensin I complex subunit G

(*NCAPG*)_rs2074974, *LCORL*_rs7700107, a disintegrin and metalloproteinase with thrombospondin motifs-like-3 (*ADAMTSL3*)_rs1600640, and insulin-like growth factor-1 receptor (*IGF1R*)_rs2871865 had an inverse association with height whereas latent-transforming growth factor beta-binding protein-1 (*LTBP1*)_rs4630744, DIS3-like exonuclease 2 (*DIS3L2*)_rs1249260, *ZBTB38*_rs6762722, and *GDF5*_rs224331 were positively associated with height (Table 3). The adjusted ORs of the genetic variants were between 0.8078 and 1.191, indicating a small effect, although each showed a highly significant *p*-value (5×10^{-7}) in the city hospital-based cohort. However, their significance level in the Ansan/Ansung plus rural cohorts was 0.02–6.8×10^{-5}. *GDF5*_rs224331 was located in the exon and was a missense mutation. These results indicated that the selected genetic variants were weakly associated with adult height (Table 3).

Table 3. Characteristics of genetic variants related to adult height from GMDR analysis.

CHR [1]	SNP [2]	Base Pair	A1 [3]	A2 [4]	OR [5]	SE [6]	*p* for City [7]	*p* for Asan + Nong [8]	MAF [9]	*p* for HWE [10]	Gene Names	Location
2	rs4630744	33461375	G	A	1.105	0.01906	1.53×10^{-7}	0.02458	0.3792	0.4099	*LTBP1*	Intron
2	rs13034890	71430542	T	C	0.9086	0.01886	3.72×10^{-7}	0.0014	0.4567	0.6295	*PAIP2B*	Intron
2	rs1249260	233046182	C	T	1.172	0.01877	2.99×10^{-17}	0.00016	0.4515	0.7705	*DIS3L2*	Downstream
3	rs6762722	141145216	G	A	1.177	0.02095	6.65×10^{-15}	0.000749	0.2546	0.5644	*ZBTB38*	Intron
4	rs3756173	8598698	T	C	0.9058	0.01917	2.48×10^{-7}	0.000515	0.4135	0.5123	*CPZ*	Intron
4	rs2074974	17812615	C	A	0.8989	0.01884	1.57×10^{-8}	0.00103	0.461	0.8746	*NCAPG*	5′ UTR
4	rs7700107	17880416	C	A	0.8112	0.02421	5.34×10^{-18}	6.78×10^{-5}	0.2072	0.1078	*LCORL*	Downstream
15	rs1600640	84603034	T	G	0.8711	0.02363	5.18×10^{-9}	0.000252	0.2115	0.5692	*ADAMTSL3*	Intron
15	rs2871865	99194896	G	C	0.8078	0.03676	6.34×10^{-9}	0.00673	0.0800	0.0640	*IGF1R*	Intron
20	rs224331	34022387	A	C	1.191	0.02073	3.13×10^{-16}	0.000956	0.2683	0.8747	*GDF5*	Missense (Ala276Ser)

[1] Chromosome; [2] Single nucleotide polymorphism; [3] Minor allele; [4] Major allele; [5] Odds ratio (OR) for city cohort; [6] Standard error; [7] *p*-value for OR after adjusting for age, gender, residence area, survey year, body mass index, daily energy intake, education and income in the hospital-based cohort (case: n = 17,545; control: n = 36,283); [8] *p*-value for OR after adjusting for covariates in the Ansan/Ansung cohort (case: n = 1657; control: n = 3245); [9] Minor allele frequency; [10] Hardy–Weinberg equilibrium. *LTBP1*, latent transforming growth factor beta binding protein-1; *PAIP2B*, poly(A)-binding protein interacting protein-2B; *DIS3L2*, DIS3 like 3′-5′ exoribonuclease-2; *ZBTB38*, zinc finger and BTB domain containing 38; *CPZ*, carboxypeptidase Z; *NCAPG*, non-SMC condensin I complex subunit G; *LCORL*, ligand-dependent nuclear receptor corepressor like; *ADAMTSL3*, a disintegrin and metalloproteinase with thrombospondin motifs-3; *IGF1R*, insulin-like growth factor-1 receptor; *GDF5*, growth differentiation factor-5.

3.5. SNP–SNP Interaction by GMDR

The genetic models, including 4, 7, 8, 9, and 10 genetic variants, met the selection criteria for the optimal model. The four-SNP model included *ZBTB38*_rs6762722, *PAIP2B*_rs13034890, *CPZ*_rs3756173, and *LTBP1*_rs4630744 whereas the seven SNP model contained *ZBTB38*_rs6762722, *PAIP2B*_rs13034890, *GDF5*_rs143384, *LCORL*_rs7700107, *DIS3L2*_rs1249260, *LTBP1*_rs4630744, and *NCAPG*_rs2074974 (Supplementary Table S1). Height was 1.29 times and 1.22 times positively associated with PRS for models 4 and 7, respectively, after adjusting for covariate groups (group 1 included age, sex, weight, education, income, and place of residence; group 2 contained covariate group 1 plus energy intake, alcohol intake, physical activity, and smoking, as shown in Figure 2). Therefore, the adjusted OR of the PRS was higher than that of the single SNP model. However, the ORs of the 4-SNP and 7-SNP models were similar, and the PRS of the 4-SNP model was appropriate. The four-SNP model was, therefore, more appropriate to explain the genetic impact on height.

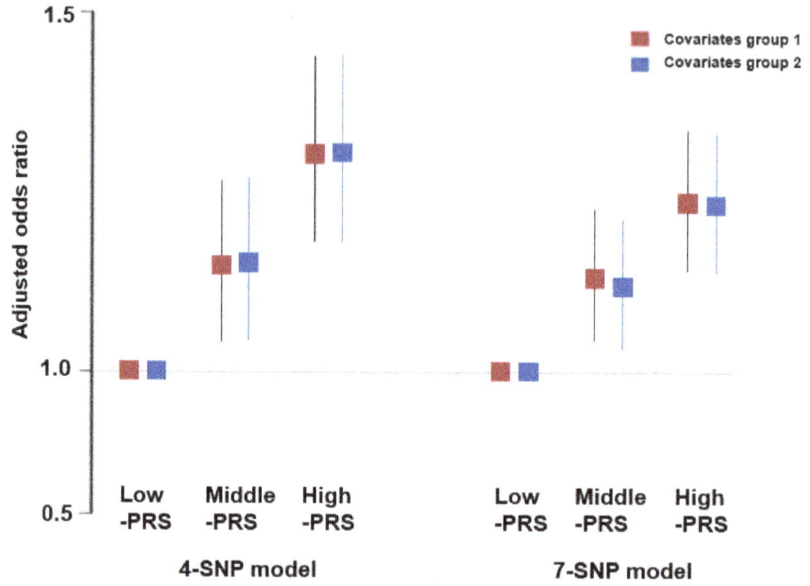

Figure 2. Adjusted odds ratio (ORs) and 95% confidence intervals (CI) of 4-SNP PRS and 7-SNP PRS models for tall stature PRS was generated with the sum of the number of risk alleles in each SNP, and it was classified as Low-PRS, Middle-PRS, and High-PRS according to the range 0–3, 4–5, and ≥6 in the four-SNP model and 0–5, 6–7, and ≥8 in the six-SNP model, respectively. Covariates of group 1 included age, gender, weight, residence area, education, and income, and those of group 2 contained those in group 1 plus energy intake, exercise, alcohol drinking, smoking, and osteoporosis incidence.

3.6. Expression of Quantitative Trait Loci (eQTL) of the Selected Genes According to the Alleles

The genetic expression of the variant alleles of *GDF5, NCAPG, LCORL, PAIP2B, ADAMTSL3*, and *IGF1R* was different across different tissues. The tibial nerve showed the maximum differences in the gene expression of the alleles. The gene expression of *NCAPG*_rs2074974 decreased with the risk allele (A) compared to the non-risk allele in the tibial nerve (slope = 0.14, p = 0.003), and *IGFR1* also showed a similar trend in the tibial nerve (slope = 0.11, p = 0.011) (Figure 3). However, *ADAMTSL3*_rs1600640 showed an increased expression with the risk allele compared to the non-risk allele in the tibial nerve (slope = 0.16, p = 0.0015) and the arterial appendage of the heart (slope = 0.16, $p = 2.1 \times 10^{-7}$). The risk allele of *PAIP2B*_rs13034890 had a lower expression than the non-risk allele in the pituitary (slope = −0.29, p = 0.000083). Interestingly, the expression of *GFD5*_rs224331 showed a significant decrement in the risk allele compared to the non-risk allele in various tissues, especially the cortex, hippocampus, and amygdala of the brain, tibial nerve, thyroid, and heart (slope = −0.12~−0.41, $p = 0.0095–1.1 \times 10^{-8}$) (Figure 3). These results indicated that the gene expressions in the tibial nerve could act as central modulators of height and that the selected genetic variants in the tibial nerve play a critical role in height.

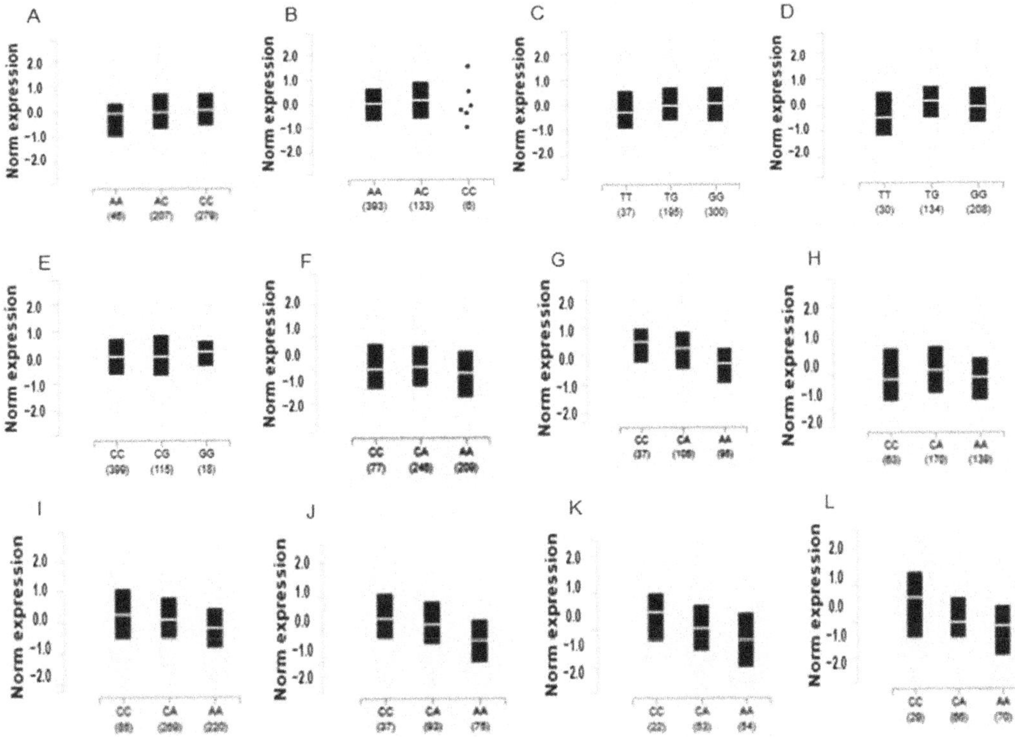

Figure 3. Gene expression according to the alleles of the selected SNPs for high body fat risk in different tissues. The black box indicated the box plot of norm expression of the gene according to each allele. The middle line was the median of the norm expression of gene with each allele. The green bell shape showed the distribution of norm gene expression with each allele. (**A**) *NCAPG*_rs2074974 in the tibial nerve (β = 0.14; *p* = 0.003). (**B**) *LCORL*_rs7700107 in the tibial nerve (β = 0.091; *p* = 0.0092). (**C**) *ADAMTSL3*_rs1600640 in the tibial tissue (β = 0.12, *p* = 0.0015). (**D**) *ADAMTSL3*_rs1600640 in the atrial appendage of the heart (β = 0.16, $p = 2.1 \times 10^{-7}$). (**E**) *IGF-1R*_rs2871865 in the tibial tissue (β = 0.11, *p* = 0.011). (**F**) *GDF5*_rs224331 in the tibial tissue (β = −0.13, *p* = 0.0095). (**G**) *GDF5*_rs224331 in the pituitary (β = −0.42, $p = 4.6 \times 10^{-12}$). (**H**) *GDF5*_rs224331 in the atrial appendage of the heart (β = −0.14, *p* = 0.0071). (**I**) *GDF5*_rs224331 in the thyroid (β = −0.25, $p = 1.1 \times 10^{-8}$). (**J**) *GDF5*_rs224331 in the cortex of the brain (β = −0.41, $p = 8.2 \times 10^{-6}$). (**K**) *GDF5*_rs224331 in the hippocampus of the brain (β = −0.38, *p* = 0.000049). (**L**) *GDF5*_rs224331 in the amygdala of the brain (β = −0.33, *p* = 0.003).

3.7. Binding Affinity of Hydrolyzable Tannins to GDF5_rs224331

The wild and mutated *GDF5*_rs224331 exhibited somewhat different binding free energy to food agents (Table 4). Among them, rugosin A is shown in Figure 4 as an example. The wild type of *GDF5*_rs224331 had a decreased binding free energy (<−10.7 kcal/mol) to hydrolyzable tannins such as stachyurin, lambertianin B, sanguiin H6, lambertianin A, mongolicain A, casuariin, punicacortein D, rugosin A, rugosin E, valolaginic acid, rugosin D, cinnamtannin II, eugenigrandin A, rugosin A, Chinese tannin, and gemin D (Table 4). However, some hydrolyzable tannins, including rugosin A, rugosin D, rugosin E, and valolaginic acid, increased binding free energy to the mutated type, indicating that they had a lower binding affinity to mutated *GDF5*. The binding energy between the wild-type GDF5 protein and rugosin A with hydrogen bond in Figure 4A and the pink and green parts indicated hydrogen donor and acceptor, respectively. Their binding position and intermolecular force are shown in the two-dimensional picture in Figure 4B. The binding and interactions of binding affinity between rugosin A and the mutated type of *GDF5* are also shown in Figure 4C,D.

Table 4. Biding energy of hydrolyzable tannins to GDF5 wild type and mutated one in rs224331 (unit: kcal/mol).

Compounds	Wide Type	Mutated Type
Stachyurin	−13.7	−13.8
Lambertianin B	−13.3	−13.3
Sanguiin H6	−13.2	−13.3
Lambertianin A	−13.2	−13.3
Mongolicain A	−12.9	−12.3
Casuariin	−12.7	−12.7
Punicacortein D	−12.5	−11.9
Rugosin E	−12.2	−5
Valolaginic acid	−12.1	−9.7
Rugosin D	−12	−5
Cinnamtannin II	−11.8	−11.8
Eugenigrandin A	−11.7	−11.8
Rugosin A	−11	−5.1
Chinese tannin	−10.9	−11
Gemin D	−10.7	−10.6

Figure 4E,F showed the root mean square deviation (RMSD) and root mean square fluctuation (RMSF) for *GDF5* wild and mutated types binding to rugosin A. RMSD for *GDF5* wild type binding with rugosin A was sustained close to 3 Å during 100 nanoseconds. RMSF for *GDF5* wild-type binding with rugosin A also did not exceed 3 nm, except at the 580 residue index in the RMSF graph. These results suggest that rugosin A stably bound to the *GDF5* wild type.

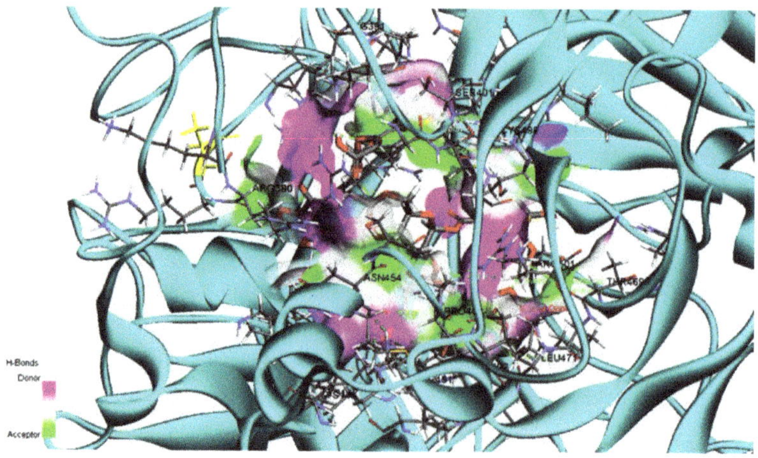

(A)

Figure 4. *Cont.*

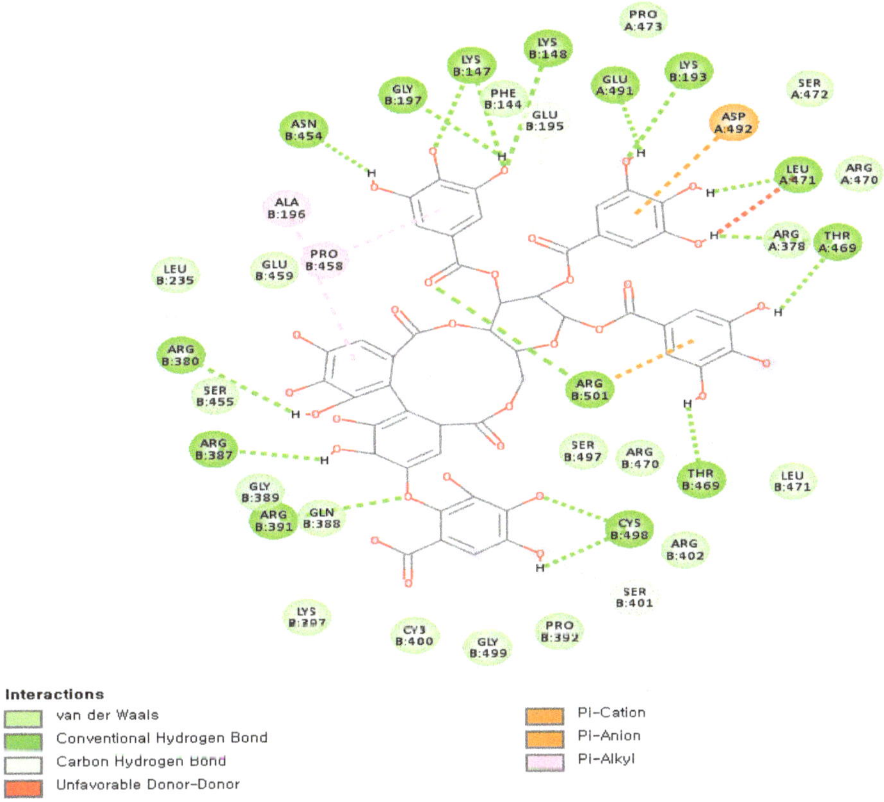

(**B**)

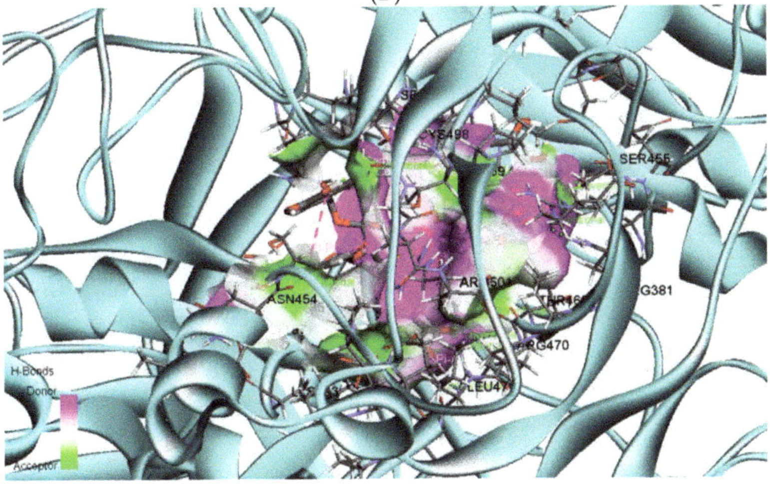

(**C**)

Figure 4. *Cont.*

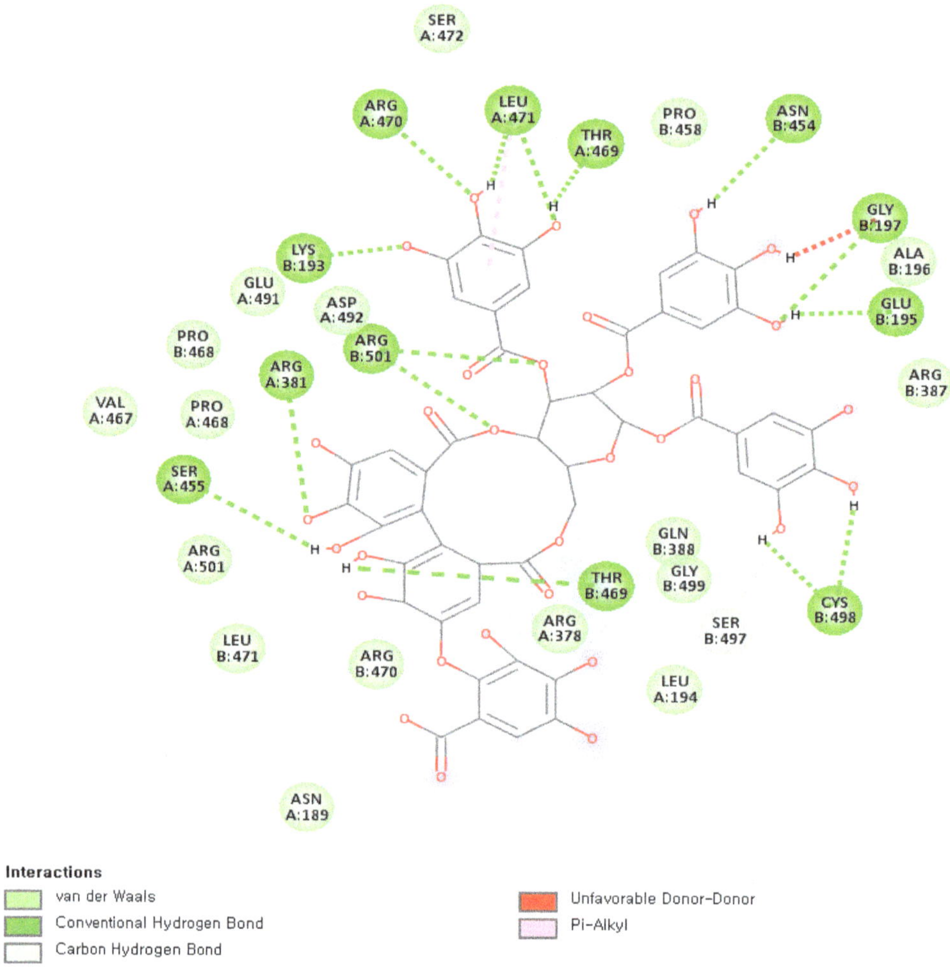

(D)

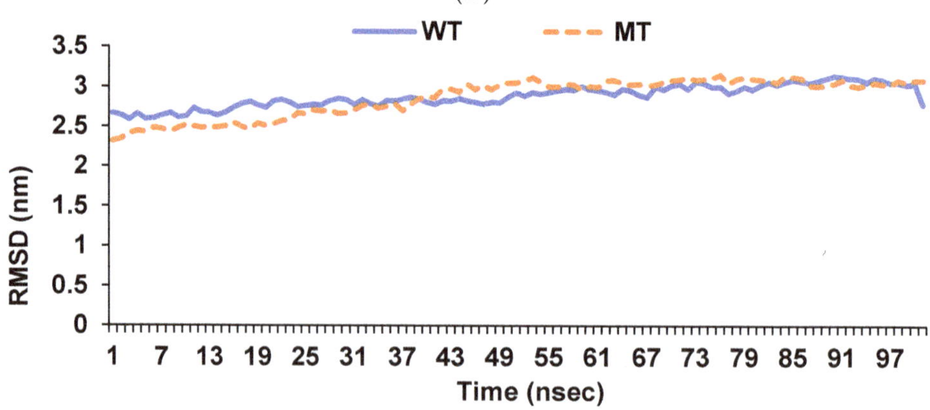

(E)

Figure 4. *Cont.*

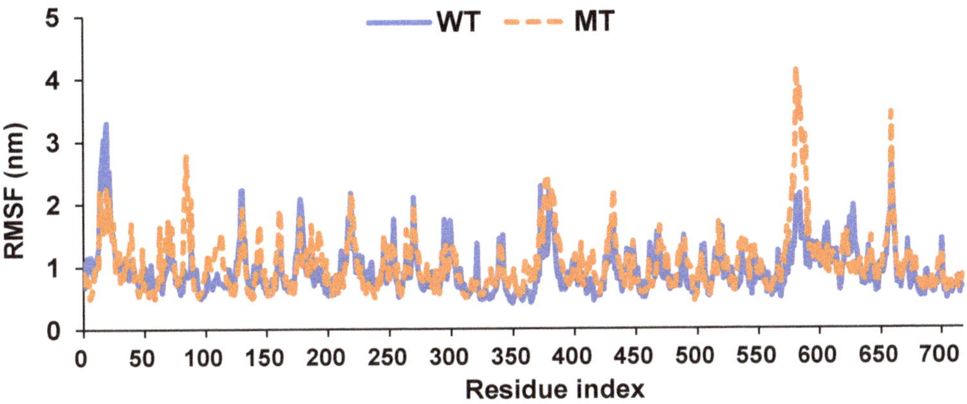

(F)

Figure 4. Molecular docking and molecular dynamic simulation (MDS) of rugosin A on growth differentiation factor-5 (*GDF-5*) rs224331 (Ala276Ser) wild and mutated types. (**A**) Molecular docking of rugosin A on *GDF5* rs224331 wild type. (**B**) The interaction force between rugosin A on *GDF5* rs224331 wild type. (**C**) Molecular docking of rugosin A on *GDF5* rs224331 mutated type. (**D**) The interaction force between rugosin A on *GDF5* rs224331 mutated type. (**E**) RMSD of rugosin A on *GDF5* rs224331 wild and mutated types. (**F**) RMSF of rugosin A on *GDF5* rs224331 wild and mutated types.

3.8. Association of PRS with the Risk of Metabolic Syndrome and Its Components

Height was greater with PRS in the ascending order. The PRS for the four-SNP model was not related to body composition, including BMI, waist circumferences (Table 5), SMI, and fat mass, except height. Osteoporosis and arthritis showed no significant change with changes in PRS. However, PRS was inversely related to WBC, an immunity index (Table 5). The PRS was inversely linked to MetS, blood HbA1c, and serum glucose concentrations, but was not significantly associated with insulin resistance. However, lipid profiles, blood pressure, and serum ALT and AST concentrations were not associated with PRS (Table 5). Therefore, the association of height-related PRS with glucose metabolism is likely related to insulin secretion. However, serum HDL, LDL, and triglyceride concentrations did not differ among the PRS groups. SBP and DBP showed a similar trend (Table 5). However, serum ALT concentrations but not AST were lower in the High-PRS compared to the Low-PRS groups and were inversely associated with PRS (Table 5).

Table 5. Adjusted means and odds ratio (ORs) of metabolic syndrome (metS)-related parameters according to polygenic risk score (PRS) of the four-SNP model generated from adult height.

	Low-PRS (n = 6107)	Middle-PRS (n = 29,668)	High-PRS (n = 22,926)	Adjusted ORs and 95 CI
Height (cm) [1]	160.4 ± 0.07 [c]	160.7 ± 0.03 [b]	160.9 ± 0.04 [a***]	1.293 (1.127–1.381)
BMI (kg/m^2) [2]	23.8 ± 0.04	23.8 ± 0.02	23.9 ± 0.02	1.054 (0.987–1.124)
Waist (cm) [3]	80.6 ± 0.10	80.7 ± 0.05	80.7 ± 0.05	0.975 (0.883–1.077)
Weight at 18 [4]	54.8 ± 0.11 [b]	55.2 ± 0.05 [a]	55.1 ± 0.06 [ab]	1.059 (0.994–1.129)
SMI [5]	7.05 ± 0.008 [a]	7.01 ± 0.004 [b]	6.98 ± 0.004 [c***]	0.960 (0.896–1.029)
Fat mass [6]	28.3 ± 0.04	28.3 ± 0.02	28.4 ± 0.02	0.959 (0.896–1.027)
WBC (10^9/L) [7]	5.78 ± 0.02 [a]	5.70 ± 0.01 [b]	5.68 ± 0.01 [b**]	0.894 (0.837–0.954)
Serum hs-CRP (mg/dL) [8]	0.152 ± 0.006 [a]	0.136 ± 0.003 [b]	0.141 ± 0.003 [ab*]	0.862 (0.675–1.100)
MetS [9]	883 (14.5)	4162 (14.0)	3255 (14.2)	0.894 (0.815–0.982)
Serum glucose (mg/dL) [10]	95.5 ± 0.26	95.1 ± 0.12	95.0 ± 0.14	0.905 (0.828–0.990)
Blood HbA1c (%) [11]	5.73 ± 0.01 [a]	5.71 ± 0.01 [b]	5.71 ± 0.01 [b*]	0.851 (0.740–0.980)
Insulin resistance (%)	1955 (11.3)	347 (12.0)	2066 (6.1)	0.953 (0.854–1.064)
Serum total cholesterol (mg/dL) [12]	197 ± 0.48	197 ± 0.22	197 ± 0.25	0.940 (0.875–1.010)

Table 5. *Cont.*

	Low-PRS (n = 6107)	Middle-PRS (n = 29,668)	High-PRS (n = 22,926)	Adjusted ORs and 95 CI
Serum HDL (mg/dL) [13]	53.6 ± 0.17	53.8 ± 0.08	53.8 ± 0.09	1.047 (0.978–1.120)
Serum LDL (mg/dL) [14]	119 ± 0.44	119 ± 0.20	119 ± 0.23	0.960 (0.883–1.043)
Serum TG (mg/dL) [15]	126 ± 1.11	125 ± 0.50	124 ± 0.57	0.981 (0.917–1.048)
SBP (mmHg) [16]	122 ± 0.19	122 ± 0.08	123 ± 0.10	1.050 (0.984–1.121)
DBP (mmHg) [17]	75.6 ± 0.12	75.6 ± 0.06	75.8 ± 0.06	1.020 (0.919–1.132)
Serum AST (IU/L) [18]	24.7 ± 0.31 [a]	23.7 ± 0.14 [b]	23.5 ± 0.16 [b]	0.875 (0.755–1.014)
Serum ALT (IU/L) [19]	23.4 ± 0.30 [a]	22.4 ± 0.14 [b]	22.1 ± 0.16 [b]	0.893 (0.812–0.981)
Egfr [20]	86.5 ± 0.20	86.0 ± 0.09	86.1 ± 0.11	1.059 (0.893–1.257)
Arthritis (N, %)	549 (9.0)	2619 (8.84)	1962 (8.57)	0.920 (0.825–1.026)
Osteoporosis (N, Yes%)	344 (5.64)	1568 (5.29)	1162 (5.08)	0.920 (0.800–1.057)

CI, confidence interval; SBP, systolic blood pressure; DBP, diastolic blood pressure; AST, aspartate aminotransferase; ALT, alanine aminotransferase; TG, triglyceride; N, the number of participants. The high-height group was ≥175 for men and ≥163 for women. Adjusting for covariates including age, gender, weight, body weight at age 18, education, income, total activity, energy intake, alcohol intake, residence area, and smoking status. PRS with four SNPs of the best GMDR model was divided into three categories according to the number of the risk alleles: ≤3, 4–5, and ≥6 into Low-PRS, Middle-PRS, and High-PRS, respectively. The cutoff points of the reference for logistic regression were as following: [1] <172.5 cm for men and <160 cm for women; [2] <25 kg/m^2 for BMI; [3] <90 cm for men and 85 cm for women waist circumferences; [4] <60 kg for men and <59 kg for women; [5] <7.6 kg/m in men and 5.4 kg/m in women for SMI in skeletal muscle index (SMI defined as appendicular skeletal muscle mass/height); [6] <25% for men and 30% for women for fat mass; [7] <4.0 × 10^9 counts/L WBC; [8] <0.5 mg/dL serum high-sensitive C-reactive protein (hs-CRP) concentrations; [9] metabolic syndrome; [10] <126 mL/dL fasting serum glucose plus diabetic drug intake; [11] <6.5% HbA1c plus diabetic drug intake; [12] <230 mg/dL plasma total cholesterol concentrations; [13] >40 mg/dL for men and 50 mg/dL for women plasma HDL cholesterol; [14] <160 mg/dL plasma total cholesterol concentrations; [15] <150 mg/dL plasma triglyceride concentrations; [16] <140 mmHg SBP, [17] <90 mmHg DBP plus hypertension medication; [18] <40 U/L aspartate aminotransferase; [19] <35 U/L alanine aminotransferase; [20] <70 mL/min/1.73 m^2 estimated glomerular filtration rate (eGFR). * Significant differences by PRS at $p < 0.05$, ** at $p < 0.01$, *** $p < 0.001$. [a, b, c] Different superscript letters indicated significant differences among the groups in Tukey's test at $p < 0.05$.

3.9. Interaction between PRS and Lifestyle Factors for Adult Height

Among the lifestyle factors, energy intake and rice-main diet interacted with PRS to influence height ($p = 0.0078$ and $p = 0.0095$) (Table 6). In people with high energy intake, PRS was positively related to height by 1.414 times but not in those with low energy intake. PRS interacted with the rice-main diet to affect height among the dietary patterns (Table 6). In adults with a low rice-main diet, PRS was positively linked to height by 1.318 times but not in those with a high rice-main diet. The results suggested that the genetic impact for adult height was not shown to be related to low energy intake, mainly containing rice.

Table 6. Adjusted odds ratios for adult height risk by polygenetic risk scores (PRS) of the 4-SNP model for SNP-SNP interaction after covariate adjustments according to the patterns of lifestyles.

	Low-PRS (n = 14,420)	Middle-PRS (n = 21,641)	High-PRS (n = 4201)	Gene-Nutrient Interaction *p* Value
Low energy [1] High energy	1	1.032 (0.914–1.165) 1.253 (1.075–1.461)	1.130 (0.999–1.278) 1.414 (1.210–1.652)	0.0078
Low KBD [2] High KBD	1	1.092 (0.971–1.228) 1.158 (0.984–1.362)	1.183 (1.050–1.333) 1.334 (1.132–1.574)	0.0923
Low PBD [2] High PBD	1	1.056 (0.938–1.188) 1.220 (1.039–1.432)	1.157 (1.026–1.305) 1.377 (1.170–1.620)	0.0567
Low WSD [2] High WSD	1	1.070 (0.937–1.222) 1.163 (1.015–1.332)	1.188 (1.038–1.359) 1.282 (1.117–1.472)	0.1898

Table 6. *Cont.*

	Low-PRS (n = 14,420)	Middle-PRS (n = 21,641)	High-PRS (n = 4201)	Gene-Nutrient Interaction p Value
Low RMD [2] High RMD	1	1.157 (1.027–1.303) 1.041 (0.888–1.220)	1.318 (1.168–1.486) 1.092 (0.929–1.284)	0.0095
Low alcohol [3] High alcohol	1	1.143 (1.003–1.303) 1.081 (0.941–1.241)	1.286 (1.125–1.469) 1.175 (1.021–1.353)	0.6960
Low exercise [4] High exercise	1	1.037 (0.901–1.194) 1.173 (1.031–1.335)	1.202 (1.042–1.386) 1.253 (1.099–1.429)	0.2233
Non-smoking Former smoking +smoking	1	1.124 (1.003–1.260) 1.095 (0.920–1.302)	1.225 (1.091–1.375) 1.260 (1.057–1.502)	0.0547

Values represent adjusted odd ratios and 95% confidence intervals. Covariates included age, sex, education, income, energy intake (percentage of estimated energy requirement), residence areas, daily activity, alcohol intake, and smoking status. PRS with four SNPs of the best GMDR model was divided into three categories according to the number of the risk alleles: ≤3, 4–5, and ≥6 into Low-PRS, Middle-PRS, and High-PRS, respectively. [1] <Estimated energy requirement defined in dietary reference index; [2] <75th percentiles; [3] <20g daily alcohol intake; [4] <moderate exercise for 150 min/day.

4. Discussion

The adult height peaked in the 20s and gradually reduced after 30 years of age. Although the loss of height varies from individual to individual, various factors are involved in the decrement. The impact of genetics explains about 10% of the height variation among people of East Asian, Hispanic, African, and South Asian ancestry [30]. An average adult loses about two centimeters in height every decade after age 30. Women tend to lose more height than men due to an accelerated rate of bone loss during menopause [31]. The loss of height can also be accelerated by certain health conditions such as osteoporosis, poor nutrition, lack of exercise, and some medical therapies such as chemotherapy and radiation therapy [32]. Therefore, the present study explored the genetic variants for tall stature using adult height and adjusting for covariates related to metabolic syndrome. Our results indicated that the selected genetic variants could explain the genetic impact on tall stature in Asians.

Short height in adults is linked to an increased risk of MetS. Previous studies have demonstrated that being short is associated with higher body fat, higher triglycerides, lower serum HDL cholesterol concentrations, and an increased risk of hypertension and type 2 diabetes, all of which are components of MetS [33,34]. However, tall stature is associated with obesity during childhood [35]. The association between stature and MetS and the impact of genetics remains unclear. The present study exhibited that tall stature was inversely linked to MetS, its components, CVD, fat and skeletal muscle mass, serum ALT and AST concentrations, and arthritis, but not osteoporosis, after adjusting height-related parameters. It was related to increased insulin resistance, which might be a primary factor for the inverse relationship between tall stature and the risk of MetS and CVD. An earlier German study demonstrated that an additional height of 10 cm is associated with a 41% and 33% lower risk of type 2 diabetes among men and women, respectively. Moreover, the association of height with a lower risk of type 2 diabetes was also seen in overweight or obese men and women (36% in men and 30% in women), although the decrease in risk was lower than that seen in normal-weight adults [33]. Therefore, increased insulin resistance in short stature elevated MetS risk, which was consistent with the results of the current study.

Common genetic variants significantly influence height. Genetic variants affect the trait by the cumulative effects of many alleles at multiple loci rather than a single genetic variant. In European ancestries, one of the most commonly identified variants is the rs1042725 variant in the *HMGA2* gene, which is associated with a 0.4 cm increase in height [36,37]. However, the present study did not include *HMGA2*_rs1042725 as a height-related genetic variant in Asians, indicating that Asians have different genetic variants for height traits. In the genetic investigation of the anthropometric traits (GIANT) consortium, 12,111 common SNPs are associated with the height trait, accounting for 10–40% of all height variations depending on the person's ancestry [30]. Previous studies have shown that genetic variants for the height trait are somewhat different between Asians and Caucasians. The genetic variants in the loci of *LCORL*, *CABLES1*, *CDK10*, *ZBTB38*, *ZNF638*, and *TSEN15* are linked to stature in Han Chinese from the Beijing study [9]. The polygenic loci of LCORL, *DIS3L2*, *EFEMP1*, *ZBTB38*, high mobility group AT-hook 1 (*HMGA1*), citrate synthase (*CS*), and *GDF5* are also shown to be linked to stature in Taiwanese [38] and Japanese [39]. Some genetic variants related to height in Asians (*ADAMTSL3*, *ZBTB38*, *LCORL*, *DIS3L2*, and *GDF5*) overlapped with those in the present study. Therefore, polygenic variants linked to height vary according to ethnicity.

Height is inversely associated with insulin resistance, and the waist-to-height ratio is used to predict insulin resistance [40]. The present study also demonstrated that height and waist-to-height ratio were inversely associated with insulin resistance. Some polygenic variants related to height, such as *IGF1R*, *GDF5*, *DIS3L2*, and *ADAMTSL3*, have been reported to modulate serum glucose and insulin concentrations, potentially [41,42]. Interestingly, the PRS of the common polygenic variants related to height was associated with MetS and its components, fasting serum glucose and blood HbA1c concentrations, but not insulin resistance in the present study. The PRS was also associated with MetS, a consequence of insulin resistance. However, PRS was not associated with insulin resistance. PRS could be associated with insufficient insulin secretion, commonly seen in Asians with type 2 diabetes, and it suggests that height-related genetic variants could be associated with insulin secretion.

Genetic variants affect gene function by modifying gene expression and/or resulting in catalytic activity by altering protein conformation, especially in missense mutations. Gene expression having risk alleles is different in different tissue types. In the present study, the polygenic variants selected for stature were expressed mainly in the tibial nerves. In the tibial nerve, the risk alleles of *NCAPG*, *LCORL*, *IGF1*, and *GDF5* had a lower expression than the non-risk allele, while the risk allele of *ADAMTSL3* showed a higher expression than the non-risk allele. *NCAPG* is responsible for condensing and stabilizing chromosomes during mitosis and meiosis. It is also involved in the carcinogenesis and progression of tumors [43]. *LCORL* is a transcription factor involved in spermatogenesis related to skeletal frame size and adult height [44,45]. The *NCAPG-LCORL* locus is associated with body growth and feed intake in cattle [46]. The IGF1R $\rightarrow$ insulin/phosphatidylinositol-3 kinase (PI3K) $\rightarrow$ protein kinase B (Akt) signaling pathway is a critical pathway for the growth of long bones [47]. The mutation in c.926C > T of *IGF1R* is involved in severe short stature in Chinese [48]. However, it did not show a significant relationship to adult stature in the present study. On the other hand, the rs2871865 variant in the intron of the *IGF1R* was significantly associated with stature. Its expression of the risk allele was lower than that of the non-risk allele in the tibial nerve. These results suggest that the genetic variations for height were mainly expressed in the tibial nerve, which receives the message from the brain for movement of the legs, feet, and toes. The tibial nerve conduction velocity is reported to be inversely associated with height [49].

Among the selected genetic variants for height, the expression of *GDF5* was the most prominent and has been revealed to be involved in height and osteoarthritis. *GDF5* variants are expressed in tissues such as the tibial nerve, brain, pituitary, thyroid, and adipose tissue. Its risk allele lowered the GDF5 protein expression more than the non-risk allele [50–52]. In addition, *GDF5*_rs224331 could affect the catalytic activity since the SNP site is a missense mutation. Hydrolyzable tannins are reported to increase the expression of collagen I, a primary component of the extracellular matrix found in the bones, to improve bone growth and ameliorate osteoarthritis [53]. They also stimulate osteoblast proliferation and differentiation and suppress osteoclast activity [54,55]. Hydrolyzable tannins have been demonstrated to have anti-inflammatory and antioxidant properties, which may help in bone growth [56]. It suggests hydrolyzable tannins may affect GDF5 to modulate bone growth and osteoarthritis. The present study demonstrated that some bioactive compounds had a decreased binding free energy with the *GDF5*_rs224331 wild type. Hydrolyzable tannins such as stachyurin, lambertianin A and B, sanguiin H6, mongolicain A, casuariin, punicacortein D, cinnamtannin II, eugenigrandin A, Chinese tannin, and gemin D had a lower binding free energy. However, the mutated *GDF5* had a decreased binding affinity (increased binding free energy) with rugosin E, rugosin A, rugosin D, and valolaginic acid. It indicated that *GDF5*_rs224331 might achieve its activity by modifying binding free energy. Bioactive compound effects can be altered with the *GDF5*_rs224331 mutation. Further experimental studies are needed to confirm this.

The study is novel in showing that polygenic variants are involved in height through SNP–SNP interactions in Asians and were also associated with immunity and glucose metabolism. Interestingly, their expressions were mainly linked to the tibial nerve. *GDF5*_rs224331 is a missense mutation; its binding affinity to some hydrolyzable tannins, such as rugosin E, rugosin A, rugosin D, and valolaginic acid, was lower in the wild type than the mutated one. Although further studies are needed, it can be suggested that bone growth is related to modulating the expression and binding affinity of *GDF5*_rs224331. However, the limitations of the present study were as follows: First, adult height was used for estimating genetic impact, which could provide good results since it was adjusted for the covariates affecting reduction in height after age 30. Second, it was conducted as a cross-sectional study; although the sample size was large (n = 58,701), the specialists gathered the samples uniformly from the volunteers. Third, data on usual food intake were gathered using the SQFFQ, in which usual food intake could be underestimated or overestimated, although it was designed for Koreans and validated with 3-day food records for four seasons.

5. Conclusions

The genetic impact on tall stature was found to be 1.29 times with the four-SNP model, including *ZBTB38*_rs6762722, *PAIP2B*_rs13034890, *CPZ*_rs3756173, and *LTBP1*_rs4630744. Although the *GDF*_rs224331 and *IGF-1R*_rs2871865 were significantly associated with tall stature, they did not interact with other SNPs to lower the adjusted OR in model 7. Furthermore, the PRS was inversely associated with MetS, hyperglycemia, and WBC risk. The SNPs in the model were explicitly expressed in the tibial nerve, which is associated with increased height. Hydrolyzable tannins lowered the binding free energy with the wild type of *GDF5*_rs143384. However, some hydrolyzable tannins (rugosin E, rugosin A, rugosin D, and valolaginic acid) did not decrease binding free energy with the mutated gene. Therefore, the genetic variants for tall stature may modulate not only height growth but also MetS, glucose metabolism, and immunity by altering the gene expression and/or their activity. Adults with PRS for short-stature need to be more cautious of the MetS risk. These results can be used in precision nutrition after further clinical study has been conducted.

Supplementary Materials: The following supporting information can be downloaded at: https://www.mdpi.com/article/10.3390/nu15071764/s1. Table S1: Generalized multifactor dimensionality reduction (GMDR) results of multi-locus interaction with genes related to adult height; Figure S1: Distribution of genetic variants for tall stature by a genome-wide association study.

Funding: This research was supported by the National Research Foundation of Republic of Korea (2019R1A2C1007203).

Institutional Review Board Statement: The institutional review board (IRB) of the Korea National Institute of Health approved the KoGES (KBP-2015-055), and the IRB of Hoseo University approved the present study (HR-034-01).

Informed Consent Statement: All participants signed a written informed consent form.

Data Availability Statement: The raw data involved in this study will be available by the corresponding author to any qualified researcher.

Conflicts of Interest: The authors declare no conflict of interest.

References

1. Zoledziewska, M.; Sidore, C.; Chiang, C.W.K.; Sanna, S.; Mulas, A.; Steri, M.; Busonero, F.; Marcus, J.H.; Marongiu, M.; Maschio, A.; et al. Height-reducing variants and selection for short stature in Sardinia. *Nat. Genet.* **2015**, *47*, 1352–1356. [CrossRef] [PubMed]
2. Huang, W.; Lei, X.; Ridder, G.; Strauss, J.; Zhao, Y. Health, Height, Height Shrinkage, and SES at Older Ages: Evidence from China. *Am. Econ. J. Appl. Econ.* **2013**, *5*, 86–121. [CrossRef] [PubMed]
3. Winkler, T.W.; Day, F.R.; Croteau-Chonka, D.C.; Wood, A.R.; Locke, A.E.; Mägi, R.; Ferreira, T.; Fall, T.; Graff, M.; Justice, A.E.; et al. Quality control and conduct of genome-wide association meta-analyses. *Nat. Protoc.* **2014**, *9*, 1192–1212. [CrossRef] [PubMed]
4. Kang, H.M.; Sul, J.H.; Service, S.K.; Zaitlen, N.A.; Kong, S.Y.; Freimer, N.B.; Sabatti, C.; Eskin, E. Variance component model to account for sample structure in genome-wide association studies. *Nat. Genet.* **2010**, *42*, 348–354. [CrossRef] [PubMed]
5. Akiyama, M.; Ishigaki, K.; Sakaue, S.; Momozawa, Y.; Horikoshi, M.; Hirata, M.; Matsuda, K.; Ikegawa, S.; Takahashi, A.; Kanai, M.; et al. Characterizing rare and low-frequency height-associated variants in the Japanese population. *Nat. Commun.* **2019**, *10*, 4393. [CrossRef]
6. Benonisdottir, S.; Oddsson, A.; Helgason, A.; Kristjansson, R.P.; Sveinbjornsson, G.; Oskarsdottir, A.; Thorleifsson, G.; Davidsson, O.B.; Arnadottir, G.A.; Sulem, G.; et al. Epigenetic and genetic components of height regulation. *Nat. Commun.* **2016**, *7*, 13490. [CrossRef]
7. Kim, J.J.; Lee, H.I.; Park, T.; Kim, K.; Lee, J.E.; Cho, N.H.; Shin, C.; Cho, Y.S.; Lee, J.Y.; Han, B.G.; et al. Identification of 15 loci influencing height in a Korean population. *J. Hum. Genet.* **2010**, *55*, 27–31. [CrossRef]
8. Zhao, J.; Li, M.; Bradfield, J.P.; Zhang, H.; Mentch, F.D.; Wang, K.; Sleiman, P.M.; Kim, C.E.; Glessner, J.T.; Hou, C.; et al. The role of height-associated loci identified in genome-wide association studies in the determination of pediatric stature. *BMC Med. Genet.* **2010**, *11*, 96. [CrossRef]
9. Lin, Y.-J.; Liao, W.-L.; Wang, C.-H.; Tsai, L.-P.; Tang, C.-H.; Chen, C.-H.; Wu, J.-Y.; Liang, W.-M.; Hsieh, A.-R.; Cheng, C.-F.; et al. Association of human height-related genetic variants with familial short stature in Han Chinese in Taiwan. *Sci. Rep.* **2017**, *7*, 6372. [CrossRef]
10. Rehunen, S.K.J.; Kautiainen, H.; Eriksson, J.G.; Korhonen, P.E. Adult height and glucose tolerance: A re-appraisal of the importance of body mass index. *Diabet. Med.* **2017**, *34*, 1129–1135. [CrossRef]
11. Rosenberg, M.A.; Kaplan, R.C.; Siscovick, D.S.; Psaty, B.M.; Heckbert, S.R.; Newton-Cheh, C.; Mukamal, K.J. Genetic variants related to height and risk of atrial fibrillation: The cardiovascular health study. *Am. J. Epidemiol.* **2014**, *180*, 215–222. [CrossRef]
12. Lai, F.Y.; Nath, M.; Hamby, S.E.; Thompson, J.R.; Nelson, C.P.; Samani, N.J. Adult height and risk of 50 diseases: A combined epidemiological and genetic analysis. *BMC Med.* **2018**, *16*, 187. [CrossRef]
13. Ryu, J.; Lee, C. Association of glycosylated hemoglobin with the gene encoding CDKAL1 in the Korean Association Resource (KARE) study. *Hum. Mutat.* **2012**, *33*, 655–659. [CrossRef]
14. Wu, X.; Park, S. An Inverse Relation between Hyperglycemia and Skeletal Muscle Mass Predicted by Using a Machine Learning Approach in Middle-Aged and Older Adults in Large Cohorts. *J. Clin. Med.* **2021**, *10*, 2133. [CrossRef]
15. Park, S.; Yang, H.J.; Kim, M.J.; Hur, H.J.; Kim, S.H.; Kim, M.S. Interactions between Polygenic Risk Scores, Dietary Pattern, and Menarche Age with the Obesity Risk in a Large Hospital-Based Cohort. *Nutrients* **2021**, *13*, 3772. [CrossRef]
16. Stevens, L.A.; Coresh, J.; Feldman, H.I.; Greene, T.; Lash, J.P.; Nelson, R.G.; Rahman, M.; Deysher, A.E.; Zhang, Y.L.; Schmid, C.H.; et al. Evaluation of the modification of diet in renal disease study equation in a large, diverse population. *J. Am. Soc. Nephrol.* **2007**, *18*, 2749–2757. [CrossRef]
17. Park, S.; Kim, C.; Wu, X. Development and Validation of an Insulin Resistance Predicting Model Using a Machine-Learning Approach in a Population-Based Cohort in Korea. *Diagnostics* **2022**, *12*, 212. [CrossRef]
18. van Woudenbergh, G.J.; Theofylaktopoulou, D.; Kuijsten, A.; Ferreira, I.; van Greevenbroek, M.M.; van der Kallen, C.J.; Schalkwijk, C.G.; Stehouwer, C.D.; Ocké, M.C.; Nijpels, G.; et al. Adapted dietary inflammatory index and its association with a summary score for low-grade inflammation and markers of glucose metabolism: The Cohort study on Diabetes and Atherosclerosis Maastricht (CODAM) and the Hoorn study. *Am. J. Clin. Nutr.* **2013**, *98*, 1533–1542. [CrossRef]
19. Lee, S.K.; Kim, M.K. Relationship of sodium intake with obesity among Korean children and adolescents: Korea National Health and Nutrition Examination Survey. *Br. J. Nutr.* **2016**, *115*, 834–841. [CrossRef]
20. Kim, Y.; Han, B.-G.; KoGES Group. Cohort Profile: The Korean Genome and Epidemiology Study (KoGES) Consortium. *Int. J. Epidemiol.* **2017**, *46*, e20. [CrossRef]
21. Rabbee, N.; Speed, T.P. A genotype calling algorithm for affymetrix SNP arrays. *Bioinformatics* **2006**, *22*, 7–12. [CrossRef] [PubMed]
22. Uma Jyothi, K.; Reddy, B.M. Gene-gene and gene-environment interactions in the etiology of type 2 diabetes mellitus in the population of Hyderabad, India. *Meta Gene* **2015**, *5*, 9–20. [CrossRef] [PubMed]
23. Khera, A.V.; Chaffin, M.; Wade, K.H.; Zahid, S.; Brancale, J.; Xia, R.; Distefano, M.; Senol-Cosar, O.; Haas, M.E.; Bick, A.; et al. Polygenic Prediction of Weight and Obesity Trajectories from Birth to Adulthood. *Cell* **2019**, *177*, 587–596. [CrossRef] [PubMed]
24. Lewis, C.M.; Vassos, E. Polygenic risk scores: From research tools to clinical instruments. *Genome Med.* **2020**, *12*, 44. [CrossRef] [PubMed]
25. Zhou, J.-Y.; Song, M.Y.; Park, S. Carbohydrate and sodium intake and physical activity interact with genetic risk scores of four genetic variants mainly related to lipid metabolism to modulate metabolic syndrome risk in Korean middle-aged adults. *Br. J. Nutr.* **2019**, *122*, 919–927. [CrossRef] [PubMed]

26. Yuan, H.; Liu, L.; Zhou, J.; Zhang, T.; Daily, J.W.; Park, S. Bioactive Components of Houttuynia cordata Thunb and Their Potential Mechanisms Against COVID-19 Using Network Pharmacology and Molecular Docking Approaches. *J. Med. Food* **2022**, *25*, 355–366. [CrossRef]
27. Yang, Y.; Shi, C.-Y.; Xie, J.; Dai, J.-H.; He, S.-L.; Tian, Y. Identification of potential dipeptidyl peptidase (DPP)-IV inhibitors among Moringa oleifera phytochemicals by virtual screening, molecular docking analysis, ADME/T-based prediction, and in vitro analyses. *Molecules* **2020**, *25*, 189. [CrossRef]
28. Committee of SAS. *SAS/STAT®14.2 User's Guide*; SAS Inc.: Cary, NC, USA, 2016.
29. Kwon, O.; Kim, H.; Kim, J.; Hwang, J.-Y.; Lee, J.; Yoon, M.O. The development of the 2020 Dietary Reference Intakes for Korean population: Lessons and challenges. *J. Nutr. Health* **2021**, *54*, 425. [CrossRef]
30. Yengo, L.; Vedantam, S.; Marouli, E.; Sidorenko, J.; Bartell, E.; Sakaue, S.; Graff, M.; Eliasen, A.U.; Jiang, Y.; Raghavan, S.; et al. A saturated map of common genetic variants associated with human height. *Nature* **2022**, *610*, 704–712. [CrossRef]
31. Sözen, T.; Özışık, L.; Başaran, N. An overview and management of osteoporosis. *Eur. J. Rheumatol.* **2017**, *4*, 46–56. [CrossRef]
32. Curtis, E.; Litwic, A.; Cooper, C.; Dennison, E. Determinants of Muscle and Bone Aging. *J. Cell. Physiol.* **2015**, *230*, 2618–2625. [CrossRef]
33. Wittenbecher, C.; Kuxhaus, O.; Boeing, H.; Stefan, N.; Schulze, M.B. Associations of short stature and components of height with incidence of type 2 diabetes: Mediating effects of cardiometabolic risk factors. *Diabetologia* **2019**, *62*, 2211–2221. [CrossRef]
34. Toro-Huamanchumo, C.J.; Pérez-Zavala, M.; Urrunaga-Pastor, D.; De La Fuente-Carmelino, L.; Benites-Zapata, V.A. Relationship between the short stature and the prevalence of metabolic syndrome and insulin resistance markers in workers of a private educational institution in Peru. *Diabetol. Metab. Syndr.* **2020**, *14*, 1339–1345. [CrossRef]
35. Chung, S. Growth and Puberty in Obese Children and Implications of Body Composition. *J. Obes. Metab. Syndr.* **2017**, *26*, 243–250. [CrossRef]
36. Yang, T.L.; Guo, Y.; Zhang, L.S.; Tian, Q.; Yan, H.; Guo, Y.F.; Deng, H.W. HMGA2 is confirmed to be associated with human adult height. *Ann. Hum. Genet.* **2010**, *74*, 11–16. [CrossRef]
37. Hendriks, A.E.; Brown, M.R.; Boot, A.M.; Oostra, B.A.; Drop, S.L.; Parks, J.S. Genetic variation in candidate genes like the HMGA2 gene in the extremely tall. *Horm. Res. Paediatr.* **2011**, *76*, 307–313. [CrossRef]
38. Chiou, J.-S.; Cheng, C.-F.; Liang, W.-M.; Chou, C.-H.; Wang, C.-H.; Lin, W.-D.; Chiu, M.-L.; Cheng, W.-C.; Lin, C.-W.; Lin, T.-H.; et al. Your height affects your health: Genetic determinants and health-related outcomes in Taiwan. *BMC Med.* **2022**, *20*, 250. [CrossRef]
39. Okada, Y.; Kamatani, Y.; Takahashi, A.; Matsuda, K.; Hosono, N.; Ohmiya, H.; Daigo, Y.; Yamamoto, K.; Kubo, M.; Nakamura, Y.; et al. A genome-wide association study in 19 633 Japanese subjects identified LHX3-QSOX2 and IGF1 as adult height loci. *Hum. Mol. Genet.* **2010**, *19*, 2303–2312. [CrossRef]
40. Lechner, K.; Lechner, B.; Crispin, A.; Schwarz, P.E.H.; von Bibra, H. Waist-to-height ratio and metabolic phenotype compared to the Matsuda index for the prediction of insulin resistance. *Sci. Rep.* **2021**, *11*, 8224. [CrossRef]
41. Cubbon, R.M.; Kearney, M.T.; Wheatcroft, S.B. Endothelial IGF-1 Receptor Signalling in Diabetes and Insulin Resistance. *Trends Endocrinol. Metab.* **2016**, *27*, 96–104. [CrossRef]
42. Zhang, W.; Wu, X.; Pei, Z.; Kiess, W.; Yang, Y.; Xu, Y.; Chang, Z.; Wu, J.; Sun, C.; Luo, F. GDF5 Promotes White Adipose Tissue Thermogenesis via p38 MAPK Signaling Pathway. *DNA Cell Biol.* **2019**, *38*, 1303–1312. [CrossRef] [PubMed]
43. Xiao, C.; Gong, J.; Jie, Y.; Cao, J.; Chen, Z.; Li, R.; Chong, Y.; Hu, B.; Zhang, Q. NCAPG Is a Promising Therapeutic Target Across Different Tumor Types. *Front. Pharmacol.* **2020**, *11*, 387. [CrossRef] [PubMed]
44. Soranzo, N.; Rivadeneira, F.; Chinappen-Horsley, U.; Malkina, I.; Richards, J.B.; Hammond, N.; Stolk, L.; Nica, A.; Inouye, M.; Hofman, A.; et al. Meta-analysis of genome-wide scans for human adult stature identifies novel Loci and associations with measures of skeletal frame size. *PLoS Genet.* **2009**, *5*, e1000445. [CrossRef] [PubMed]
45. Metzger, J.; Schrimpf, R.; Philipp, U.; Distl, O. Expression Levels of LCORL Are Associated with Body Size in Horses. *PLoS ONE* **2013**, *8*, e56497. [CrossRef] [PubMed]
46. Lindholm-Perry, A.K.; Sexten, A.K.; Kuehn, L.A.; Smith, T.P.; King, D.A.; Shackelford, S.D.; Wheeler, T.L.; Ferrell, C.L.; Jenkins, T.G.; Snelling, W.M.; et al. Association, effects and validation of polymorphisms within the NCAPG—LCORL locus located on BTA6 with feed intake, gain, meat and carcass traits in beef cattle. *BMC Genet.* **2011**, *12*, 103. [CrossRef] [PubMed]
47. Sergi, C.; Shen, F.; Liu, S.M. Insulin/IGF-1R, SIRT1, and FOXOs Pathways-An Intriguing Interaction Platform for Bone and Osteosarcoma. *Front. Endocrinol.* **2019**, *10*, 93. [CrossRef]
48. Yang, Y.; Huang, H.; Chen, K.; Yang, L.; Xie, L.L.; Xiong, T.; Wu, X. Novel mutation of type-1 insulin-like growth factor receptor (IGF-1R) gene in a severe short stature pedigree identified by targeted next-generation sequencing. *J. Genet.* **2019**, *98*, 16. [CrossRef]
49. Campbell, W.W., Jr.; Ward, L.C.; Swift, T.R. Nerve conduction velocity varies inversely with height. *Muscle Nerve* **1981**, *4*, 520–523. [CrossRef]
50. Lettre, G. The osteoarthritis and height GDF5 locus yields its secrets. *Nat. Genet.* **2017**, *49*, 1165–1166. [CrossRef]
51. Sanna, S.; Jackson, A.U.; Nagaraja, R.; Willer, C.J.; Chen, W.M.; Bonnycastle, L.L.; Shen, H.; Timpson, N.; Lettre, G.; Usala, G.; et al. Common variants in the GDF5-UQCC region are associated with variation in human height. *Nat. Genet.* **2008**, *40*, 198–203. [CrossRef]

52. Wu, D.D.; Li, G.M.; Jin, W.; Li, Y.; Zhang, Y.P. Positive selection on the osteoarthritis-risk and decreased-height associated variants at the GDF5 gene in East Asians. *PLoS ONE* **2012**, *7*, e42553. [CrossRef]
53. Ekambaram, S.P.; Perumal, S.S.; Erusappan, T.; Srinivasan, A. Hydrolysable tannin-rich fraction from Terminalia chebula Retz. fruits ameliorates collagen-induced arthritis in BALB/c mice. *Inflammopharmacol* **2020**, *28*, 275–287. [CrossRef]
54. Li, K.; Zhang, X.; He, B.; Yang, R.; Zhang, Y.; Shen, Z.; Chen, P.; Du, W. Geraniin promotes osteoblast proliferation and differentiation via the activation of Wnt/β-catenin pathway. *Biomed. Pharmacother.* **2018**, *99*, 319–324. [CrossRef]
55. Kim, S.-Y.; Park, J.-Y.; Lee, S.H.; Woo, J.T.; Kima, T.H.; Park, E.K. Furosin, an ellagitannin, suppresses RANKL-induced osteoclast differentiation and function through inhibition of MAP kinase activation and actin ring formation. *Biochem. Biophys. Res. Commun.* **2004**, *325*, 1471–1480.
56. Tong, Z.; He, W.; Fan, X.; Guo, A. Biological Function of Plant Tannin and Its Application in Animal Health. *Front. Vet. Sci.* **2022**, *8*, 803657. [CrossRef]

nutrients

MDPI

Article

Interaction between the *PNPLA3* Gene and Nutritional Factors on NAFLD Development: The Korean Genome and Epidemiology Study

Sooyeon Oh [1], Jooho Lee [2,*], Sukyung Chun [2], Ja-Eun Choi [3], Mi Na Kim [2], Young Eun Chon [2], Yeonjung Ha [2], Seong-Gyu Hwang [2], Sang-Woon Choi [1] and Kyung-Won Hong [3,*]

1 Chaum Life Center, CHA University School of Medicine, Seoul 06062, Republic of Korea
2 Department of Gastroenterology, CHA Bundang Medical Center, CHA University School of Medicine, Seongnam 13496, Republic of Korea
3 Healthcare R&D Division, Theragen Bio Co., Ltd., Suwon 16229, Republic of Korea
* Correspondence: ljh0505@cha.ac.kr (J.L.); kyungwon.hong@theragenbio.com (K.-W.H.); Tel.: +82-31-780-1811 (J.L.); +82-10-3922-7089 (K.-W.H.); Fax: +82-31-780-1812 (J.L.); +82-31-888-9440 (K.-W.H.)

Abstract: Genetic and nutritional factors contribute to the development of non-alcoholic fatty liver disease (NAFLD); however, gene–diet interactions in NAFLD development are poorly understood. In this case–control study, a large dataset from the Korean Genome and Epidemiology Study cohort (n = 72,299) comprising genomic data, medical records, social history, and dietary data was used. We investigated the interactions between the *PNPLA3* rs738409 genotype and nutritional factors and their possible effect on the risk of NAFLD development in 2950 patients with NAFLD and 12,907 controls. In the *PNPLA3* risk allele group, high protein, fat, sodium, phosphorus, niacin, and vitamin B6 intakes were associated with a decreased risk of NAFLD. In the non-risk allele group, only high fat intake was associated with a decreased risk of NAFLD. Among these nutrients, high sodium intake had a significant protective interaction with the *PNPLA3* genotype against NAFLD (p = 0.002). Among salty foods, only kimchi had a significant protective effect against the *PNPLA3* genotype (p = 0.012). Thus, the *PNPLA3* genotype is differentially associated with nutritional factors. In particular, it interacts with kimchi, a fermented vegetable dish. Therefore, fermented vegetables may serve as a tailored therapeutic food for people with the *PNPLA3* risk allele.

Keywords: non-alcoholic fatty liver disease; sodium; kimchi; fermented vegetable

1. Introduction

Non-alcoholic fatty liver disease (NAFLD) is one of the most common liver diseases worldwide, and its prevalence continues to increase [1]. As a spectrum of liver disease that begins with simple steatosis, NAFLD is considered a hepatic manifestation of metabolic syndrome. Owing to an increase in obesity, the total number of NAFLD cases is projected to increase by 0–30% between 2016–2030 [2]. This is a public health concern because NAFLD can progress to non-alcoholic steatohepatitis, cirrhosis, and ultimately hepatocellular carcinoma (HCC). Statistics have shown that the incidence of HCC associated with NAFLD is increasing [3]. If this trend persists, NAFLD will become the principal cause of liver transplantation and liver-related mortality in the coming decades [4].

Traditionally, a sedentary lifestyle with a high calorie intake but low energy output has been considered the major cause of hepatic fat accumulation. Recent investigations have provided new insights into the pathogenesis of NAFLD. Genome-wide association studies (GWAS) have discovered that some single nucleotide polymorphisms (SNPs) increase the risk of NAFLD, including patatin-like phospholipase domain-containing 3 (*PNPLA3*), transmembrane 6 superfamily member 2 (*TM6SF2*), sorting and assembly machinery component 50 (*SAMM-50*), farnesyl diphosphate farnesyl transferase I (*FDFT1*), collagen type XIII alpha 1 (*COL13A1*), neurocan (*NCAN*), glucokinase regulatory protein (*GCKR*), membrane-bound O-acyltransferase domain-containing protein 7 (*MBOAT7*), apolipoprotein C3 (*APOC3*), sterol regulatory element binding transcription factor 2 (*SREBF2*) rs133291, membrane-bound O-acyltransferase domain-containing 7 transmembrane channel-like 4 (*MBOAT7-TMC4*) rs641738, 17β-hydroxysteroid dehydrogenase type 13 (*HSD17B13*), and serpin family member 1 (*SERPINA1*) [5–9]. Among these, the association between *PNPLA3* and NAFLD risk is the most robustly observed [6]. The *PNPLA3* rs738409 C > G SNP, which leads to the replacement of isoleucine with methionine at position 148 (I148M), contributes to the development of NAFLD by promoting triglyceride synthesis and accumulation in hepatocytes [10]. The effect of this SNP was even observed in subjects with "lean NAFLD", a condition where non-obese individuals are predisposed to develop NAFLD [11].In these circumstances, where genetic factors directly affect the pathogenesis but no specific drug therapy exists, the role of diet becomes crucial. Thus far, only a few studies have suggested the role of gene–diet interactions in NAFLD development. One such study discovered that a high intake of carbohydrates, isoflavones, and methionine increased the risk of hepatic fibrosis in NAFLD in a *PNPLA3* genotype-dependent manner [12]. Another study showed that the minor allele of the haplotype in the 22q13 loci increased the risk of NAFLD via interaction with a high carbohydrate intake, including high consumption of noodles and meat [13]. Therefore, certain dietary patterns should be avoided by individuals with risk genotypes, and certain dietary patterns could potentially attenuate the genetic influence. These findings justify the idea of precision nutrition, a diet tailored to one's genetic make-up [10,12–15]. In addition, dietary factors modulate the crosstalk between the gut microbiome and the liver through portal circulation [16,17]. Therefore, diet can also affect the gut-liver axis that plays a critical role in the development of NAFLD.In Korea, the prevalence of NAFLD (32.87–42.9%) exceeded the global average (25.24%) [9,18]. It is expected to increase in the future owing to a Westernized diet, lack of exercise, and increase in obesity and type 2 diabetes [9]. The Korean diet is based on steamed rice served with small side dishes; these side dishes are prepared mainly with vegetables, and less frequently with meat, poultry, or fish. Fermented vegetables such as kimchi are also used quite frequently. Furthermore, the traditional Korean diet uses whole mixed grains and beans; thus, it is known to be healthy, having a low glycemic index, low cholesterol, and high fiber content [19]. However, these traditional benefits have been overshadowed by an excessive consumption of refined rice, along with an increased consumption of Western foods [20].

In this study, we hypothesized that a distinctive gene–diet interaction may exist between *PNPLA3* and NAFLD. We planned to test this interaction in the Korean population. The majority of Koreans belong to one ethnicity, called han minjok [21], which results in a genetic and dietary homogeneity that is beneficial for genetic and dietary studies. We used a large Korean cohort (Korean Genome and Epidemiology Study [KoGES] cohort) dataset obtained from the Korea Biobank Project (KBP) database.

2. Materials and Methods

2.1. Study Population and the Definition of NAFLD

In this case–control study, we received a dataset from the KoGES City cohort (n = 72,299) from KBP [22] that was collected from January 2004 to December 2012. The dataset included genomic data, medical records, social history, and dietary data. Patients with NAFLD were screened for the case group if they had a medical record of fatty liver disease but had neither viral hepatitis nor alcohol consumption exceeding 210 g or more per week for men and 140 g or more per week for women. Subjects without NAFLD or other underlying diseases were screened for the control group. Subjects missing medical records of fatty liver disease were excluded.

2.2. Dietary Assessment

For each nutrient, a daily intake either above or below the recommended daily value was defined as high intake or low intake, respectively. The recommended daily values used in this study are presented in Supplementary Table S1. The median value adjusted by the total energy was taken as the cut-off, and a higher intake above the median was defined as a high intake value.

For the dietary assessment, a semi-quantitative food-frequency questionnaire (FFQ) [23–25] comprising 103 items was developed for the KoGES data [26]. Participants reported the frequency and number of foods consumed over the past year using the FFQ. The intake of a food item three times or more per week was defined as high intake and less than three times a week as low intake. The KoGES data also provided information on the intake of 23 nutrients.

2.3. Genome-Wide Genotyping

The genotype data were provided by the Center for Genome Science, Korea National Institute of Health and were processed using the Korea Biobank Array (Affymetrix, Santa Clara, CA, USA). The experimental results of the Korea Biobank Array were filtered using the following quality control criteria: call rate > 97%, minor allele frequency > 1%, and Hardy–Weinberg equilibrium, $p < 1 \times 10^{-5}$. After quality control filtering, the experimental phenotypes were used to analyze the genotype datasets from the 1000 Genome Phase 1 and 2 Asian panels. GWAS identified 7,975,321 SNPs on chromosomes 1–22. The SNP genotype (*PNPLA3* rs738409) of participants was extracted from the Korea Biobank array (referred to as KoreanChip), which was optimized for the Korean population to demonstrate the findings of blood biochemical traits through GWAS [27].

2.4. Statistical Analysis

We performed GWAS in our KoGES City cohort (n = 72,299) to identify the genetic indicators of NAFLD. The genetic risk of the *PNPLA3* rs738409 SNP for NAFLD development was tested using additive, dominant, and recessive models. We compared the baseline characteristics between participants with and without NAFLD using the Student's t-test for continuous variables and Pearson's chi-squared test for categorical variables. Furthermore, the association between the *PNPLA3* rs738409 SNP and NAFLD was assessed using Pearson's chi-squared test and logistic regression analysis. The association was adjusted for age, sex, body mass index, smoking status, and alcohol intake. The influence of dietary factors on the *PNPLA3* rs738409 SNP risk allele and non-risk allele groups was investigated, as was the interaction between dietary factors and the *PNPLA3* rs738409 SNP. All genetic association tests were conducted using PLINK version 1.9 (https://www.cog-ge-nom-ics.org/plink) [28], and the phenotype characteristics were analyzed using the SPSS (IBM SPSS Statistics Inc., New York, NY, USA) [29] and R statistical software (v4.1.2; R Core Team) [30]. Differences were considered statistically significant at two-sided p values < 0.05.

3. Results

In the KoGES City cohort (n = 72,299), the *PNPLA3* rs738409 *G* allele frequency was 0.41. Among the subjects (n = 72,299), those with missing data (n = 10,676) were excluded. In a review of medical records, subjects with fatty liver disease but without viral hepatitis were screened (n = 3568). Among them, participants with alcohol consumption exceeding the predefined criteria were excluded (n = 618). Among the controls (n = 58,055), subjects with underlying diseases were excluded (n = 45,148). Additionally, we excluded subjects with missing nutrient intake data in both the control group (n = 105) and in patients with NAFLD (n = 27). Finally, 2950 patients with NAFLD and 12,907 healthy controls were included in the analyses (Figure 1). The baseline characteristics of the patients with NAFLD and healthy controls are summarized in Table 1. The NAFLD group exhibited more elements of metabolic syndrome, such as higher blood pressure, higher serum levels of glucose and cholesterol, and a higher waist circumference. In contrast, there were more current drinkers and smokers in the control group (both $p < 0.001$) than in the NAFLD group.

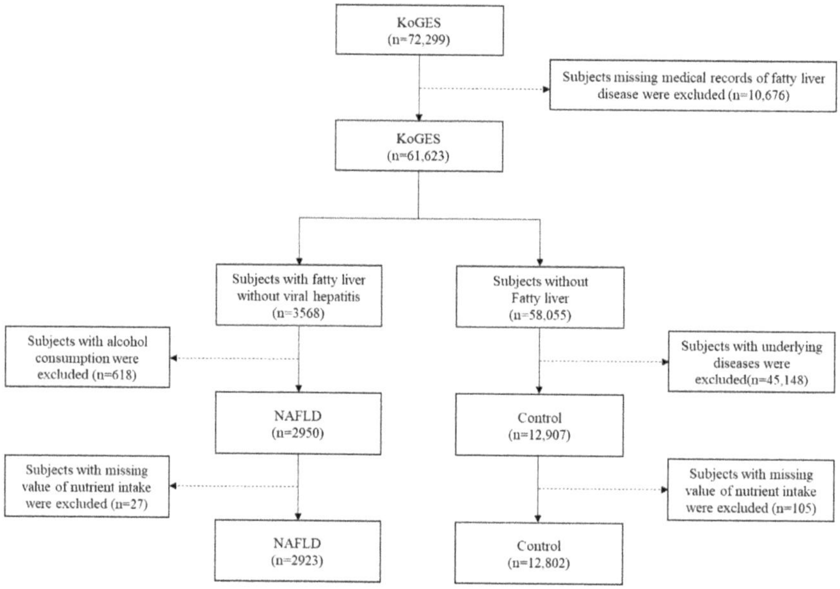

Figure 1. Study population using KoGES database. Abbreviations: KoGES, Korean Genome and Epidemiology Study; NAFLD, non-alcoholic fatty liver disease.

Table 1. Baseline characteristics of the study population.

Characteristics	Total Population (n = 15,725)	NAFLD (n = 2923)	Control (n = 12,802)	p-Value
Age (years)	49.84 (±7.82)	55.77 (±7.58)	48.48 (±7.22)	0.01
Male (%)	7108 (45.20)	1481 (50.68)	5630 (43.98)	<0.001
Female (%)	8617 (54.80)	1442 (49.32)	7172 (56.02)	
Glucose (mg/dL)	92.09 (±15.03)	100.67 (±23.30)	90.12 (±11.51)	<0.001
Total cholesterol (mg/dL)	193.17 (±33.09)	197.11 (±37.00)	192.27 (±32.06)	<0.001
HDL cholesterol (mg/dL)	53.73 (±12.98)	48.49 (±11.37)	54.93 (±13.03)	<0.001
Triglyceride (mg/dL)	114.78 (±75.87)	153.08 (±100.12)	106.02 (±66.08)	<0.001
Waist circumference (cm)	80.26 (±8.55)	85.9 (±7.91)	78.97 (±8.16)	0.01
Systolic blood pressure (mmHg)	119.98 (±14.03)	125.68 (±14.46)	118.68 (±13.60)	<0.001
Diastolic blood pressure (mmHg)	74.84 (±9.58)	77.47 (±9.50)	74.24 (±9.50)	<0.001
Body mass index (kg/m^2)	23.67 (±2.82)	25.53 (±2.88)	23.24 (±2.62)	0.01
Alcohol intake				<0.001
Non-drinker (n (%))	6823 (43.50)	1368 (46.92)	5455 (42.70)	
Ex-drinker (n (%))	579 (3.70)	206 (7.08)	373 (2.95)	
Current drinker (n (%))	8275 (52.80)	1340 (46.00)	6935 (54.35)	
Smoking				<0.001
Non-smoker (n (%))	10,590 (67.5)	1850 (63.39)	8740 (68.43)	
Ex-smoker (n (%))	2862 (18.3)	703 (24.13)	2159 (16.90)	
Current smoker (n (%))	2238 (14.3)	364 (12.47)	1874 (14.67)	
PNPLA3 rs738409 genotype				
GG (n (%))	2911 (18.4)	652 (22.2)	2259 (17.5)	<0.001
GC (n (%))	7705 (48.6)	1458 (48.4)	6247 (48.4)	
CC (n (%))	5241 (33.0)	840 (28.4)	4401 (34.1)	

Values are mean (± standard deviation) or number (%). Abbreviations: n, number; NAFLD, non-alcoholic fatty liver disease; *PNPLA3*, patatin-like phospholipase domain-containing 3.

The frequency of the *PNPLA3* rs738409 *G* allele was 0.43 in the total included population, while it was 0.47 and 0.42 in the NAFLD and control groups, respectively. In our study, the *PNPLA3* risk allele group (rs738409, *GG* + *GC*) had a higher proportion of NAFLD cases than the non-risk allele group (rs738409, *CC*) (19.9% [2110/10,616] vs. 16.0% [840/5241]; Table 1). The *PNPLA3* rs738409 *G* allele showed a significant association with NAFLD in all additive, dominant, and recessive model analyses (odds ratio [OR] of the additive model = 1.22; 95% confidence interval, 1.15–1.30; $p = 1.96 \times 10^{-10}$) (Supplementary Table S2). After adjusting for age, sex, alcohol intake, and smoking, the *PNPLA3* rs738409 *G* allele remained an independent risk factor for NAFLD (adjusted OR = 1.22; $p = 1.75 \times 10^{-10}$) (Table 2).

Table 2. Genetic risk of *PNPLA3* rs738409 *G* allele on non-alcoholic fatty liver disease (NAFLD) occurrence in multivariable logistic regression analyses.

	NAFLD (*n*)	Control (*n*)	Proportion of NAFLD (%)	Additive Model	
				OR (95% CI)	*p*-Value *
PNPLA3 rs738409 allele					
C allele	837	4380	16.04	1	
G allele	2100	8452	19.90	1.22 (1.15~1.30)	1.75×10^{-10}
Sex					
Male	1485	5652	20.81	1	
Female	1452	7180	16.82	0.80 (0.75–0.85)	4.79×10^{-11}
Age					
<49 years old	521	7430	6.55	1	5.8×10^{-280}
>=49 years old	2416	5402	30.9	4.68 (4.28–5.11)	
Alcohol					
Non-drinker	1379	5479	20.11	1	
Ex-drinker	208	378	35.49	1.76 (1.56–1.98)	1.77×10^{-16}
Current drinker	1350	6975	16.22	0.81 (0.75–0.86)	7.26×10^{-10}
Smoking					
Non-smoker	1863	8781	17.5	1	
Ex-smoker	708	2169	24.61	1.40 (1.30–1.52)	2.9×10^{-17}
Current smoker	366	1882	16.28	0.93 (0.84–1.03)	0.177

Abbreviations: NAFLD, non-alcoholic fatty liver disease; 95% CI, 95% confidence interval; OR, odds ratio; *PNPLA3*, patatin-like phospholipase domain-containing 3. * The *p*-values were calculated with R Statistical Software (v 4.1.2; R Core Team 2021).

3.1. Association between Nutrients and NAFLD in the Total Population Subsection

In the macronutrient-consumption assessments, the healthy control group contained a significantly larger proportion of subjects with high protein ($p < 0.001$) and fat ($p < 0.001$) intakes (Supplementary Table S3) compared to the NAFLD group. The proportion of subjects with a high carbohydrate intake was larger in the NAFLD group than in the control group, but this difference was not statistically significant ($p = 0.067$) (Supplementary Table S3).

In the mineral- and vitamin-consumption assessments, a significantly larger proportion of subjects in the healthy control group had higher intakes of sodium ($p = 0.003$), phosphorus ($p < 0.001$), zinc ($p = 0.04$), vitamin B1 ($p < 0.001$), vitamin B2 ($p < 0.047$), niacin ($p < 0.001$), and vitamin B6 ($p < 0.001$) than those in the NAFLD group (Supplementary Table S3).

Subsequently, the proportions of NAFLD cases were compared between the high- and low-nutrient-intake groups (Table 3). High intakes of protein (OR = 0.765, $p < 0.0001$), fat (OR = 0.616, $p < 0.0001$), sodium (OR = 0.884, $p < 0.0001$), phosphorus (OR = 0.850, $p < 0.0001$), zinc (OR = 0.919, $p = 0.043$), vitamin B1 (OR = 0.771, $p < 0.0001$), vitamin B2 (OR = 0.874, $p = 0.044$), niacin (OR = 0.746, $p < 0.0001$), and vitamin B6 (OR =0.842, $p < 0.0001$) significantly decreased the risk of NAFLD.

Table 3. Comparison of the non-alcoholic fatty liver disease (NAFLD) case proportions between several nutrients' high-intake and low-intake groups among the * total participants (n = 15,725), as determined by univariate analyses.

Nutrient	High-Intake Group **			Low-Intake Group **			OR (95% CI)	*p*-Value
	NAFLD (n)	Control (n)	Proportion of NAFLD (%)	NAFLD (n)	Control (n)	Proportion of NAFLD (%)		
Energy	786	3823	17.05%	2137	8979	19.22%	0.864 (0.789–0.945)	0.001
Carbohydrate	2354	10,113	18.88%	569	2689	17.46%	1.10 (0.994–1.217)	0.064
Protein	1396	6971	16.68%	1527	5831	20.75%	0.765 (0.706–0.829)	<0.0001
Fat	1294	7209	15.22%	1629	5593	22.56%	0.616 (0.568–0.668)	<0.0001
Sodium	1678	7732	17.83%	1245	5070	19.71%	0.884 (0.815–0.959)	<0.0001
Potassium	285	1281	18.20%	2638	11,521	18.63%	0.972 (0.849–1.112)	0.677
Calcium	353	1508	18.97%	2570	11,294	18.54%	1.029 (0.909–1.164)	0.654
Phosphorus	2021	9282	17.88%	902	3520	20.40%	0.850 (0.778–0.927)	<0.0001
Iron	342	1460	18.98%	2581	11,342	18.54%	1.029 (0.908–1.167)	0.651
Zinc	1190	5475	17.85%	1733	7327	19.13%	0.919 (0.847–0.997)	0.043
Vitamin A	493	2160	18.58%	2430	10,642	18.59%	1.00 (0.898–1.113)	0.994
Carotene	1547	7026	18.05%	1376	5776	19.24%	0.924 (0.853–1.002)	0.055
Vitamin B1	383	2094	15.46%	2540	10,708	19.17%	0.771 (0.686–0.867)	<0.0001
Vitamin B2	302	1491	16.84%	2621	11,311	18.81%	0.874 (0.767–0.996)	0.044
Niacin	1052	5504	16.05%	1871	7298	20.41%	0.746 (0.686–0.81)	<0.0001
Vitamin B6	1553	7344	17.46%	1370	5458	20.06%	0.842 (0.777–0.913)	<0.0001
Folate	209	820	20.31%	2714	11,982	18.47%	1.125 (0.961–1.317)	0.142
Vitamin C	1265	5585	18.47%	1658	7217	18.68%	0.986 (0.909–1.069)	0.732
Vitamin E	395	1870	17.44%	2528	10,932	18.78%	0.913 (0.813–1.027)	0.129
Ash	491	2171	18.44%	2432	10,631	18.62%	0.989 (0.888–1.101)	0.835
Cholesterol	352	1715	17.03%	2571	11,087	18.82%	0.885 (0.783–1.0)	0.051

Abbreviations: NAFLD, non-alcoholic fatty liver disease; OR, odds ratio. * The proportion of NAFLD patients in the total population was 18.59% (2923/15,725). ** For each nutrient, daily intake above the recommended daily value was defined as high intake and below that as low intake. The recommended daily values used in this study are presented in Supplementary Table S1.

3.2. *Association between Nutrients and NAFLD According to the PNPLA3 Genotype Figures, Tables, and Schemes*

To investigate the differential association between nutrient intake and NAFLD according to the *PNPLA3* genotype, the population was divided into two groups (Table 4). In the *PNPLA3* risk allele group (rs738409, *GG* + *GC*), high intakes of protein (OR 0.821, p = 0.001), fat (OR 0.755, p < 0.0001), sodium (OR 0.771, p < 0.0001), phosphorus (OR 0.851, p = 0.009), niacin (OR 0.800, p < 0.0001), vitamin B6 (OR 0.823, p = 0.001), and ash (OR 0.833, p = 0.017) were associated with a decreased risk of NAFLD (Table 4). In the non-risk allele group (rs738409, *CC*), a high fat intake was associated with a decreased risk of NAFLD (OR 0.794, p = 0.009), but other nutrients did not show a significant association (Table 4).

Table 4. Association and interaction between nutrients and the *PNPLA3* genotype for the risk of non-alcoholic fatty liver disease (NAFLD) development. Multivariate logistic regression analyses adjusted for age, sex, and smoking were applied.

Food	*rs738409 GG + GC* (n = 10,616)				rs738409 CC (n = 5241)				Interaction p
	Proportion of NAFLD (%)		OR (95% CI)	p	Proportion of NAFLD (%)		OR (95% CI)	p	
	High Intake	Low Intake			High Intake	Low Intake			
Energy	18	20.6	0.946 (0.835–1.071)	0.381	15.1	16.4	0.943 (0.778–1.141)	0.551	0.931
Carbohydrate	20.3	18.3	1.037 (0.901–1.194)	0.617	16	15.8	0.94 (0.757–1.171)	0.577	0.545
Protein	17.7	22.3	0.821 (0.734–0.918)	0.001	14.8	17.5	0.863 (0.727–1.026)	0.094	0.381
Fat	16.1	24.2	0.755 (0.675–0.845)	<0.0001	13.4	19.1	0.794 (0.667–0.945)	0.009	0.154
Sodium	18.5	21.9	0.771 (0.689–0.863)	<0.0001	16.5	15.2	1.017 (0.853–1.214)	0.851	0.002
Potassium	19.8	19.9	0.991 (0.821–1.191)	0.922	15.2	16.1	0.901 (0.678–1.186)	0.466	0.508
Calcium	20.2	19.8	0.997 (0.837–1.183)	0.972	16.6	15.9	0.954 (0.736–1.226)	0.715	0.626
Phosphorus	18.9	22.3	0.851 (0.755–0.96)	0.009	15.9	16.4	0.971 (0.801–1.181)	0.77	0.11
Iron	19.4	19.9	0.98 (0.822–1.164)	0.817	18.1	15.7	1.226 (0.948–1.576)	0.115	0.284
Zinc	19	20.5	0.948 (0.847–1.061)	0.354	15.5	16.4	0.917 (0.771–1.091)	0.331	0.758
Vitamin A	19.3	20	0.905 (0.779–1.05)	0.19	17.2	15.7	1.075 (0.861–1.337)	0.517	0.208
Vitamin B1	16.1	20.6	0.881 (0.748–1.033)	0.122	14.1	16.4	0.889 (0.697–1.128)	0.339	0.801
Vitamin B2	18.2	20.1	0.959 (0.8–1.145)	0.649	14.3	16.2	0.819 (0.621–1.07)	0.15	0.696
Niacin	16.9	21.9	0.8 (0.713–0.897)	<0.0001	14.3	17.2	0.843 (0.706–1.005)	0.057	0.275
Vitamin B6	18.5	21.6	0.823 (0.736–0.92)	0.001	15.3	16.9	0.865 (0.728–1.028)	0.099	0.596
Folate	20.4	19.8	0.961 (0.765–1.2)	0.73	20.2	15.7	1.325 (0.968–1.796)	0.074	0.255
Vitamin C	19.8	20	0.97 (0.867–1.085)	0.594	15.8	16.1	1.016 (0.854–1.208)	0.859	0.862

Table 4. *Cont.*

Food	rs738409 GG + GC (n = 10,616)				rs738409 CC (n = 5241)				Interaction p
	Proportion of NAFLD (%)		OR (95% CI)	p	Proportion of NAFLD (%)		OR (95% CI)	p	
	High Intake	Low Intake			High Intake	Low Intake			
Vitamin E	18.6	20.1	1.016 (0.864–1.191)	0.849	15.1	16.2	0.985 (0.77–1.251)	0.9	0.869
Ash	19.1	20	0.833 (0.716–0.967)	0.017	17.2	15.7	1.032 (0.826–1.283)	0.782	0.449

Abbreviations: n, number; NAFLD, non-alcoholic fatty liver disease; 95% CI, 95% confidence interval; OR, odds ratio; *PNPLA3*, patatin-like phospholipase domain-containing 3.

Subsequently, the interaction between nutrients and the *PNPLA3* genotype was analyzed. Among the nutrients investigated, only a high sodium intake had a statistically significant protective interaction with the *PNPLA3* genotype against NAFLD development (interaction p = 0.002) (Table 4). The data are described in more detail in Supplementary Tables S4 and S5 and Figure 2. In these univariate analyses, a high intake of sodium exhibited a significant protective effect against NAFLD in the *PNPLA3* risk group (p < 0.0001) (Supplementary Table S4), while no such effect was observed in the *PNPLA3* non-risk group (p = 0.851) (Supplementary Table S5).

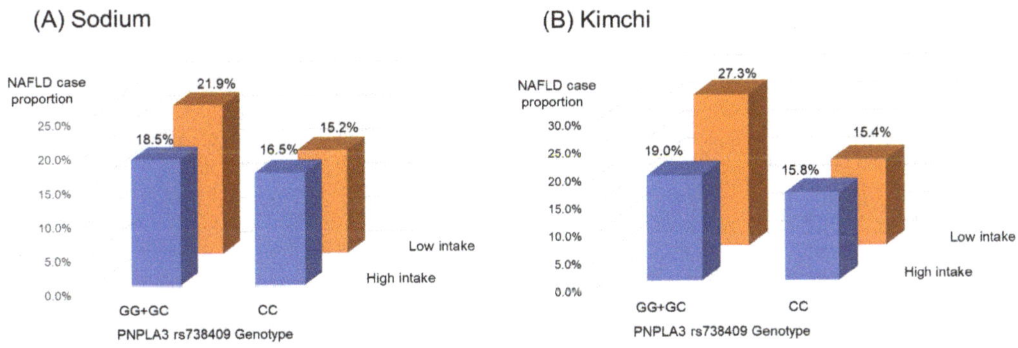

Figure 2. Impact of dietary sodium (**A**) and kimchi (**B**) intake on the genetic susceptibility of NAFLD. Abbreviations: NAFLD, non-alcoholic fatty liver disease; *PNPLA3*, patatin-like phospholipase domain-containing 3.

3.3. Protective Foods against NAFLD in the PNPLA3 Risk Allele Group

To determine which food the sodium originated from, the association between food-intake frequency and NAFLD was investigated in the *PNPLA3* genotype groups (Table 5). In the *PNPLA3* risk allele group (rs738409, *GG* + *GC*), white rice, baechu kimchi, leaf mustard or scallion kimchi, green pepper, orange, coffee, sugar (for tea/coffee), and cream (for tea/coffee) were protective against NAFLD (all p < 0.05), while multigrain rice, yogurt, nuts, pickled radish, vegetable salad, and fried food increased the risk of NAFLD development (all p < 0.05). In the *PNPLA3* non-risk allele group (rs738409, *CC*), white rice, baechu kimchi, green pepper, orange, strawberry, pear, coffee, sugar (for tea/coffee), and cream (for tea/coffee) were protective (all p < 0.05), whereas multigrain rice, yogurt, green tea, and fried food increased the NAFLD risk (all p < 0.05).

Among the foods studied, baechu kimchi had a statistically significant protective interaction with the *PNPLA3* genotype against NAFLD development (p = 0.012) (Table 5 and Figure 2). This interaction between baechu kimchi and the *PNPLA3* genotype was very similar to the interaction between sodium and the *PNPLA3* genotype. Based on these observations, we concluded that the protective effect of a high sodium intake was probably due to the intake of baechu kimchi.

Table 5. Association and interaction between foods and the *PNPLA3* genotype for the risk of non-alcoholic fatty liver disease (NAFLD) development. Multivariate logistic regression analyses adjusted for age, sex, and smoking were applied.

Food	*rs738409 GG + GC* (*n* = 10,616)				rs738409 CC (*n* = 5241)				Interaction *p*
	Proportion of NAFLD (%)		OR (95% CI)	*p*	Proportion of NAFLD (%)		OR (95% CI)	*p*	
	High Intake	Low Intake			High Intake	Low Intake			
Protective Foods									
White rice	15.07	21.13	0.785 (0.68–0.9)	0.001	10.69	17.3	0.674 (0.53–0.85)	0.001	0.689
Baechu kimchi	19.03	27.33	0.689 (0.57–0.83)	<0.001	15.79	15.4	0.745 (0.63–0.87)	<0.001	0.012
Leaf mustard or scallion kimchi	18.89	19.87	0.758 (0.62–0.94)	0.007	15.85	16	0.871 (0.71–0.99)	0.369	0.474
Green pepper	19.71	20.29	0.849 (0.73–0.99)	0.034	15.1	16.1	0.842 (0.74–0.96)	0.008	0.643
Orange	17.65	20.44	0.792 (0.66–0.95)	0.011	15.84	16.3	0.841 (0.72–0.98)	0.023	0.187
Strawberry	19.55	20.3	0.871 (0.75–1.02)	0.077	13.74	16.7	0.861 (0.76–0.98)	0.025	0.795
Pear	19.05	20.25	0.868 (0.41–0.67)	0.08	13.53	16.4	0.861 (0.47–0.71)	0.03	0.873
Coffee	17.62	26.25	0.743 (0.66–0.84)	<0.001	14.09	21.9	0.741 (0.67–0.82)	<0.001	0.487
Sugar for tea or coffee	17.64	22.68	0.751 (0.67–0.84)	<0.001	13.91	19.1	0.745 (0.68–0.82)	<0.001	0.757
Cream for tea or coffee	17.33	22.35	0.753 (0.68–0.84)	<0.001	13.8	18.1	0.761 (0.7–0.83)	<0.001	0.28
Risk Foods									
Multi grain rice	21.89	17.79	1.173 (1.06–1.3)	0.003	18.11	13.61	1.289 (1.09–1.52)	0.002	0.58
Yoghurt	23.55	19.49	1.159 (1–1.34)	0.045	19.04	15.3	1.147 (1.02–1.29)	0.025	0.59
Nuts (peanut, almond, pine nut)	26.54	19.34	1.284 (1.02–1.61)	0.032	23.02	15.44	1.297 (0.93–1.82)	0.131	0.703
Pickled radish	21.65	19.84	1.282 (1.03–1.59)	0.024	17.49	15.79	1.193 (0.85–1.67)	0.305	0.748
Vegetable salad	23.65	19.84	1.289 (1–1.66)	0.046	13.64	16.14	1.134 (0.91–1.41)	0.254	0.038
Green tea	19.56	19.89	1.135 (0.99–1.3)	0.065	17.15	15.59	1.149 (1.03–1.28)	0.014	0.554
Fried food	17.71	20.61	1.232 (1.10–1.38)	<0.001	14.34	16.54	1.254 (1.05–1.503)	0.014	0.745

Abbreviations: *n*, number; NAFLD, non-alcoholic fatty liver disease; 95% CI, 95% confidence interval; OR, odds ratio; *PNPLA3*, patatin-like phospholipase domain-containing 3.

4. Discussion

Using the KoGES database, we discovered that nutrients and foods differentially affect the risk of NAFLD depending on the *PNPLA3* genotype, and an interaction exists between nutrients or foods and the *PNPLA3* genotype in the risk of NAFLD. A high fat intake decreased the risk of NAFLD, regardless of the *PNPLA3* genotype. High intakes of protein, phosphorus, sodium, niacin, vitamin B6, and carotene were associated with a decreased risk of NAFLD in the *PNPLA3* risk allele group. Among these protective nutrients, only sodium showed a significant interaction with the *PNPLA3* genotype. To identify the source of the sodium, the association between food intake frequency and the *PNPLA3* genotype was investigated. Only baechu kimchi, a salty food, had a significant protective interaction with the *PNPLA3* genotype with respect to high sodium intake: a high intake of baechu kimchi significantly decreased the risk of NAFLD, with a greater magnitude in the *PNPLA3* risk allele group than in the *PNPLA3* non-risk allele group. Therefore, the preventive effect of a high sodium intake in the *PNPLA3* risk allele group appears to be derived from high intake of baechu kimchi.

In the current study, a high sodium intake was inversely associated with NAFLD. This observation was a conundrum. It is possible that a high-sodium diet may indeed have a preventive effect on NAFLD. A recent animal study demonstrated that mice fed a high-sodium and high-fat diet developed less hepatic steatosis, metabolic syndrome, and insulin resistance compared with those fed a normal or low-sodium and high-fat diet [31]. The reduced steatosis was associated with lower serum aldosterone levels and downregulation of hepatic mineralocorticoid receptors; thus, decreased activation of hepatic mineralocorticoids may have resulted in beneficial downstream inhibition of lipogenesis [31]. However, many studies have reported the opposite: a high-salt diet has been linked to increased glucocorticoid production, insulin resistance, metabolic syndrome, and NAFLD development [32–34]. To draw a reasonable explanation for our results, we searched for food items that correlated with a high sodium intake. Among the salty foods we investigated, salted seafood, cheese, and pickled radish increased the risk of NAFLD. Kimchi alone exerted a protective effect. Thus, we concluded that the major source of dietary sodium in this study was kimchi [35], and that the high sodium intake may have been a confounding factor.

Kimchi, a representative Korean food, is a salted and fermented vegetable usually prepared from winter cabbage [36]. The health benefits of kimchi have been widely reported [36]. Among these is an improvement in metabolic markers such as blood glucose level, cholesterol level, insulin resistance, body weight, and body mass index, which have been tested in both animals [37–39] and humans [40–42]. Furthermore, kimchi exerts a beneficial effect against hepatic damage by reducing hepatic lipid synthesis and inflammatory cytokines [39]. These beneficial effects of kimchi on metabolic markers result from their modulation of gut microbiota [43–46]. Tests performed on probiotics extracted from kimchi were found to have the same beneficial effects [45,47–49].

The key factor linking kimchi and its protective effect against NAFLD may be the short-chain fatty acids (SCFAs) produced by the gut microbiome [50]. Among SCFAs, butyrate is known to reduce obesity and related metabolic complications, including NAFLD [50,51]. An animal study demonstrated that mice fed sodium butyrate and a high-fat diet had significantly decreased hepatic steatosis and decreased hepatic triglyceride and cholesterol levels compared with mice fed only a high-fat diet [52]. Another animal study demonstrated that SCFAs (acetate, propionate, and butyrate) ameliorate methionine- and choline-deficient diet-induced hepatic steatosis and inflammation [53]. A human study tested the gut microbiome and found fewer butyrate-producing bacteria in patients with NAFLD [54]. These studies indicated that SCFAs, including butyrate, exert a preventive effect against NAFLD. Probiotics (such as *Lactobacillus casei*) combined with plant extracts reduced the markers of NAFLD and increased the concentration of butyric acid in a mouse model [55]. As a fermented vegetable, kimchi simultaneously possesses the characteristics of both probiotics and prebiotics. Thus, it enhances gut microbial diversity [43] and, in turn, has a beneficial effect on NAFLD. In particular, the protective effect of kimchi on *PNPLA3* risk genotypes indicates that fermented vegetables, such as kimchi, may be chosen as tailored foods for those affected.

In the present study, a high fat and protein intake had a preventive effect against NAFLD. A high-fat and low-carbohydrate diet, often known as a ketogenic diet, is a popular weight-reduction method. Many clinical trials have demonstrated that a ketogenic diet is effective in improving NAFLD [56–62]. A ketogenic diet decreases intrahepatic insulin resistance and, thus, serum insulin concentration [56–62]. This, in turn, increases the net hydrolysis of intrahepatic triglycerides, thus improving hepatic steatosis, inflammation, and fibrosis [56–62]. Similarly, a low-carbohydrate, high-protein diet also has metabolic benefits with regard to NAFLD. A human study demonstrated that an isocaloric low-carbohydrate diet with a high protein content decreased hepatic de novo lipogenesis, increased serum β-hydroxybutyrate concentrations (reflecting increased mitochondrial β-oxidation), and increased folate-producing *Streptococcus* and serum folate concentrations [63]. Furthermore, transcriptomic analysis revealed downregulation of the fatty acid synthesis pathway and upregulation of the folate-mediated one-carbon metabolism and fatty acid oxidation pathways [63]. Our study demonstrated that a high fat intake produces a protective effect regardless of the *PNPLA3* genotype, whereas a high intake of protein is more beneficial in the *PNPLA3* risk allele group. Nevertheless, caution should be exercised, as a ketogenic diet can elevate liver enzymes and worsen lipid profiles in patients with NAFLD [64,65]. Additionally, this diet seems less effective in premenopausal women [66]; its recommendation should therefore be carefully considered for these patients.

PNPLA3 encodes a calcium-independent triacylglycerol lipase that contains a phospholipase domain. Upon liver injury, *PNPLA3* expression is induced in hepatic stellate cells (HSCs) [67]. In the presence of *PNPLA3* I148M, HSCs become more proinflammatory and profibrogenic and produce less retinol, which potentially mediates the transition of HSCs into a myofibroblast-like phenotype [67,68]. Our findings suggest that a high retinol intake reduces the risk of NAFLD and may be relevant in this regard. Whether retinol intake prevents NAFLD in *PNPLA3* risk genotypes warrants further investigation.

This study had several limitations. First, there may have been a recall bias in collecting data on food intake habits because of the self-reporting nature of the food-frequency questionnaire. Second, the identification of fatty liver depended solely on the medical records of the subjects. The nationwide database did not include the results of imaging studies such as sonography, computed tomography, or magnetic resonance imaging. Thus, a detailed evaluation of NAFLD severity was not possible. Third, the present study could not count the various beneficial nutrients such as antioxidants, polyunsaturated fatty acids, or amino acids. As these beneficial nutrients may directly attenuate the oxidative stress that results in hepatic fibrosis [14,69], future studies need to consider these detailed aspects of nutrition.

In the present study, the associations and interactions between NAFLD, the *PNPLA3* genotype, and nutritional and dietary factors were investigated. The genetic risk of NAFLD conferred by *PNPLA3* in the Korean population was verified. In addition, we discovered some nutritional and dietary factors that can significantly influence the development of NAFLD, and these factors may interact with the *PNPLA3* genotype. Both increasing the proportion of fat and protein relative to that of carbohydrate and consuming fermented vegetables may help reduce NAFLD, regardless of the *PNPLA3* genotype. In *PNPLA3* risk genotypes, the role of a high-protein diet appears to be more important, as does meeting the daily requirement of vitamins and minerals. Furthermore, kimchi, a fermented vegetable, may induce an added protective effect in *PNPLA3* risk genotypes because of its gene–diet interaction. These results suggest that planning a therapeutic diet with high protein and fermented vegetables may be a valid strategy for those with *PNPLA3* risk genotypes and NAFLD, and warrants future interventional studies. Furthermore, in the near future, studies are needed on the mechanisms of interactions between various beneficial nutrients, *PNPLA3*, and NAFLD development.

5. Conclusions

The *PNPLA3* genotype is differentially associated with nutritional and dietary factors. In particular, it interacts with kimchi, a fermented vegetable. This indicates that tailored nutritional therapy based on an individual's genetic background may be a good strategy to prevent NAFLD, and fermented vegetables may serve as a therapeutic food for those with the *PNPLA3* risk allele.

Supplementary Materials: The following supporting information can be downloaded at: https://www.mdpi.com/article/10.3390/nu15010152/s1. Supplementary Table S1. Recommended daily intake of each nutrient according to the 2020 Korean Dietary Reference Intakes; Supplementary Table S2. Genetic risk of the *PNPLA3* rs738409 *G* allele for non-alcoholic fatty liver disease (NAFLD) occurrence in genome-wide association studies (GWAS) analyses; Supplementary Table S3. Proportion of patients with a *high intake of each nutrient; Supplementary Table S4. Non-alcoholic fatty liver disease (NAFLD) case frequencies by nutrient-consumption level in the *PNPLA3 risk group (rs738409, GG + GC) (n = 10,530) determined by univariate analyses; Supplementary Table S5. Non-alcoholic fatty liver disease (NAFLD) case frequencies by nutrient-consumption level in the *PNPLA3 non-risk group (rs738409, CC) (n = 5195) determined by univariate analyses.

Author Contributions: Conceptualization, K.-W.H. and J.L.; methodology, K.-W.H., J.L. and S.-W.C.; software, S.C. and J.-E.C.; validation, M.N.K., Y.E.C., Y.H. and S.-G.H.; formal analysis, S.C. and J.-E.C.; investigation, J.L. and S.O.; resources, K.-W.H. and J.L.; data curation, K.-W.H.; writing—original draft preparation, S.O.; writing—review and editing, M.N.K., Y.E.C., Y.H., S.-G.H., J.L. and S.-W.C.; visualization, J.L. and S.O.; supervision, S.-W.C.; project administration, J.L.; funding acquisition, K.-W.H. and J.L. All authors have read and agreed to the published version of the manuscript.

Funding: This study was conducted using bioresources from the National Biobank of Korea, the Centers for Disease Control and Prevention, Republic of Korea (2019-059). This work was supported by the Technology Innovation Program (2002781, A Platform for Prediction and Management of Health Risk Based on Personal Big Data and Lifelogging), funded by the Ministry of Trade, Industry and Energy (Republic of Korea). This study was also supported by the Research Supporting Program of the Korean Association for the Study of the Liver and the Korean Liver Foundation (2020-260).

Institutional Review Board Statement: This study was conducted in accordance with the principles described in the Declaration of Helsinki. The institutional review board of the Theragen Etex Bio Institute approved this study (internal review board no. 700062-20190819-GP-006-02).

Informed Consent Statement: Informed consent was obtained from all subjects involved in the study.

Data Availability Statement: Not applicable.

Conflicts of Interest: The authors declare no conflict of interest.

References

1. Younossi, Z.; Anstee, Q.M.; Marietti, M.; Hardy, T.; Henry, L.; Eslam, M.; George, J.; Bugianesi, E. Global burden of NAFLD and NASH: Trends, predictions, risk factors and prevention. *Nat. Rev. Gastroenterol. Hepatol.* **2018**, *15*, 11–20. [CrossRef] [PubMed]
2. Estes, C.; Anstee, Q.M.; Arias-Loste, M.T.; Bantel, H.; Bellentani, S.; Caballeria, J.; Colombo, M.; Craxi, A.; Crespo, J.; Day, C.P.; et al. Modeling NAFLD disease burden in China, France, Germany, Italy, Japan, Spain, United Kingdom, and United States for the period 2016–2030. *J. Hepatol.* **2018**, *69*, 896–904. [CrossRef] [PubMed]
3. Dyson, J.; Jaques, B.; Chattopadyhay, D.; Lochan, R.; Graham, J.; Das, D.; Aslam, T.; Patanwala, I.; Gaggar, S.; Cole, M. Hepatocellular cancer: The impact of obesity, type 2 diabetes and a multidisciplinary team. *J. Hepatol.* **2014**, *60*, 110–117. [CrossRef] [PubMed]
4. Anstee, Q.M.; Targher, G.; Day, C.P. Progression of NAFLD to diabetes mellitus, cardiovascular disease or cirrhosis. *Nat. Rev. Gastroenterol. Hepatol.* **2013**, *10*, 330–344. [CrossRef]
5. Tong, M.; Wang, F. APOC3rs2854116, PNPLA3rs738409, and TM6SF2rs58542926 polymorphisms might influence predisposition of NAFLD: A meta-analysis. *IUBMB Life* **2020**, *72*, 1757–1764. [CrossRef]
6. Lorenz, B.; Georg, S.; Hannes, O.; Stephan, Z.; Michael, S.; Leonora, D.; David, N.; Alexandra, F.; Felix, S.; Christian, D.; et al. PNPLA3 is the dominant SNP linked to liver disease severity at time of first referral to a tertiary center. *Dig. Liver Dis.* **2021**, *54*, 84–90. [CrossRef]
7. Carlsson, B.; Lindén, D.; Brolén, G.; Liljeblad, M.; Bjursell, M.; Romeo, S.; Loomba, R. Review article: The emerging role of genetics in precision medicine for patients with non-alcoholic steatohepatitis. *Aliment. Pharmacol. Ther.* **2020**, *51*, 1305–1320. [CrossRef]
8. Gao, F.; Zheng, K.I.; Chen, S.-D.; Lee, D.H.; Wu, X.-X.; Wang, X.-D.; Targher, G.; Byrne, C.D.; Chen, Y.-P.; Kim, W.; et al. Individualized Polygenic Risk Score Identifies NASH in the Eastern Asia Region: A Derivation and Validation Study. *Clin. Transl. Gastroenterol.* **2021**, *12*, e00321. [CrossRef]
9. Huh, Y.; Cho, Y.J.; Nam, G.E. Recent Epidemiology and Risk Factors of Nonalcoholic Fatty Liver Disease. *J. Obes. Metab. Syndr.* **2022**, *31*, 17–27. [CrossRef]
10. Mullins, V.A.; Bresette, W.; Johnstone, L.; Hallmark, B.; Chilton, F.H. Genomics in Personalized Nutrition: Can You "Eat for Your Genes"? *Nutrients* **2020**, *12*, 3118. [CrossRef]
11. Vilarinho, S.; Ajmera, V.; Zheng, M.; Loomba, R. Emerging Role of Genomic Analysis in Clinical Evaluation of Lean Individuals With NAFLD. *Hepatology* **2021**, *74*, 2241–2250. [CrossRef] [PubMed]
12. Vilar-Gomez, E.; Pirola, C.J.; Sookoian, S.; Wilson, L.A.; Belt, P.; Liang, T.; Liu, W.; Chalasani, N. Impact of the Association Between PNPLA3 Genetic Variation and Dietary Intake on the Risk of Significant Fibrosis in Patients With NAFLD. *Am. J. Gastroenterol.* **2021**, *116*, 994–1006. [CrossRef] [PubMed]
13. Park, S.; Kang, S. High carbohydrate and noodle/meat-rich dietary patterns interact with the minor haplotype in the 22q13 loci to increase its association with non-alcoholic fatty liver disease risk in Koreans. *Nutr. Res.* **2020**, *82*, 88–98. [CrossRef] [PubMed]
14. Hernandez-Rodas, M.C.; Valenzuela, R.; Videla, L.A. Relevant Aspects of Nutritional and Dietary Interventions in Non-Alcoholic Fatty Liver Disease. *Int. J. Mol. Sci.* **2015**, *16*, 25168–25198. [CrossRef]
15. Kalafati, I.P.; Dimitriou, M.; Borsa, D.; Vlachogiannakos, J.; Revenas, K.; Kokkinos, A.; Ladas, S.D.; Dedoussis, G.V. Fish intake interacts with TM6SF2 gene variant to affect NAFLD risk: Results of a case-control study. *Eur. J. Nutr.* **2019**, *58*, 1463–1473. [CrossRef]
16. Fianchi, F.; Liguori, A.; Gasbarrini, A.; Grieco, A.; Miele, L. Nonalcoholic Fatty Liver Disease (NAFLD) as Model of Gut-Liver Axis Interaction: From Pathophysiology to Potential Target of Treatment for Personalized Therapy. *Int. J. Mol. Sci.* **2021**, *22*, 6485. [CrossRef]
17. Kong, L.; Lu, Y.; Zhang, S.; Nan, Y.; Qiao, L. Role of nutrition, gene polymorphism, and gut microbiota in non-alcoholic fatty liver disease. *Discov. Med.* **2017**, *24*, 95–106.
18. Lee, H.W.; Kim, B.K.; Kim, S.U.; Park, J.Y.; Kim, D.Y.; Ahn, S.H.; Kim, K.J.; Han, K.H. Prevalence and Predictors of Significant Fibrosis Among Subjects with Transient Elastography-Defined Nonalcoholic Fatty Liver Disease. *Dig. Dis. Sci.* **2017**, *62*, 2150–2158. [CrossRef]
19. Kim, M.J.; Park, S.; Yang, H.J.; Shin, P.-K.; Hur, H.J.; Park, S.-J.; Lee, K.-H.; Hong, M.; Kim, J.H.; Choi, S.-W.; et al. Alleviation of Dyslipidemia via a Traditional Balanced Korean Diet Represented by a Low Glycemic and Low Cholesterol Diet in Obese Women in a Randomized Controlled Trial. *Nutrients* **2022**, *14*, 235. [CrossRef]
20. Lim, M.Y.; Hong, S.; Bang, S.J.; Chung, W.H.; Shin, J.H.; Kim, J.H.; Nam, Y.D. Gut Microbiome Structure and Association with Host Factors in a Korean Population. *mSystems* **2021**, *6*, e0017921. [CrossRef]
21. Wang, C.C.; Yeh, H.Y.; Popov, A.N.; Zhang, H.Q.; Matsumura, H.; Sirak, K.; Cheronet, O.; Kovalev, A.; Rohland, N.; Kim, A.M.; et al. Genomic insights into the formation of human populations in East Asia. *Nature* **2021**, *591*, 413–419. [CrossRef] [PubMed]
22. Kim, Y.; Han, B.G. Cohort Profile: The Korean Genome and Epidemiology Study (KoGES) Consortium. *Int. J. Epidemiol.* **2017**, *46*, 1350. [CrossRef] [PubMed]

23. Kwon, Y.J.; Kim, J.O.; Park, J.M.; Choi, J.E.; Park, D.H.; Song, Y.; Kim, S.J.; Lee, J.W.; Hong, K.W. Identification of Genetic Factors Underlying the Association between Sodium Intake Habits and Hypertension Risk. *Nutrients* **2020**, *12*, 2580. [CrossRef] [PubMed]
24. Kwon, Y.J.; Park, D.H.; Choi, J.E.; Lee, D.; Hong, K.W.; Lee, J.W. Identification of the interactions between specific genetic polymorphisms and nutrient intake associated with general and abdominal obesity in middle-aged adults. *Clin. Nutr.* **2022**, *41*, 543–551. [CrossRef] [PubMed]
25. Lee, S.B.; Choi, J.E.; Park, B.; Cha, M.Y.; Hong, K.W.; Jung, D.H. Dyslipidaemia-Genotype Interactions with Nutrient Intake and Cerebro-Cardiovascular Disease. *Biomedicines* **2022**, *10*, 1615. [CrossRef]
26. Ahn, Y.; Kwon, E.; Shim, J.E.; Park, M.K.; Joo, Y.; Kimm, K.; Park, C.; Kim, D.H. Validation and reproducibility of food frequency questionnaire for Korean genome epidemiologic study. *Eur. J. Clin. Nutr.* **2007**, *61*, 1435–1441. [CrossRef] [PubMed]
27. Moon, S.; Kim, Y.J.; Han, S.; Hwang, M.Y.; Shin, D.M.; Park, M.Y.; Lu, Y.; Yoon, K.; Jang, H.M.; Kim, Y.K.; et al. The Korea Biobank Array: Design and Identification of Coding Variants Associated with Blood Biochemical Traits. *Sci. Rep.* **2019**, *9*, 1382. [CrossRef]
28. Purcell, S.; Neale, B.; Todd-Brown, K.; Thomas, L.; Ferreira, M.A.; Bender, D.; Maller, J.; Sklar, P.; de Bakker, P.I.; Daly, M.J.; et al. PLINK: A tool set for whole-genome association and population-based linkage analyses. *Am. J. Hum. Genet.* **2007**, *81*, 559–575. [CrossRef]
29. Gray, C.D.; Kinnear, P.R. *IBM SPSS Statistics 19 Made Simple*; Psychology Press: New York, NY, USA, 2012.
30. R Core Team. *R: A Language and Environment for Statistical Computing*; R Foundation for Statistical Computing: Vienna, Austria, 2021.
31. Cabrera, D.; Rao, I.; Raasch, F.; Solis, N.; Pizarro, M.; Freire, M.; Sáenz De Urturi, D.; Ramírez, C.A.; Triantafilo, N.; León, J.; et al. Mineralocorticoid receptor modulation by dietary sodium influences NAFLD development in mice. *Ann. Hepatol.* **2021**, *24*, 100357. [CrossRef]
32. Baudrand, R.; Campino, C.; Carvajal, C.A.; Olivieri, O.; Guidi, G.; Faccini, G.; Vohringer, P.A.; Cerda, J.; Owen, G.; Kalergis, A.M.; et al. High sodium intake is associated with increased glucocorticoid production, insulin resistance and metabolic syndrome. *Clin. Endocrinol.* **2014**, *80*, 677–684. [CrossRef]
33. van den Berg, E.H.; Gruppen, E.G.; Blokzijl, H.; Bakker, S.J.L.; Dullaart, R.P.F. Higher Sodium Intake Assessed by 24 Hour Urinary Sodium Excretion Is Associated with Non-Alcoholic Fatty Liver Disease: The PREVEND Cohort Study. *J. Clin. Med.* **2019**, *8*, 2157. [CrossRef] [PubMed]
34. Kim, M.J.; Lee, K.J. Analysis of the dietary factors associated with suspected pediatric nonalcoholic fatty liver disease and potential liver fibrosis: Korean National Health and Nutrition Examination Survey 2014-2017. *BMC Pediatr.* **2020**, *20*, 121. [CrossRef] [PubMed]
35. Jeong, Y.; Kim, E.S.; Lee, J.; Kim, Y. Trends in sodium intake and major contributing food groups and dishes in Korea: The Korea National Health and Nutrition Examination Survey 2013–2017. *Nutr. Res. Pract.* **2021**, *15*, 382–395. [CrossRef]
36. Park, K.-Y.; Jeong, J.-K.; Lee, Y.-E.; Daily, J.W., III. Health benefits of kimchi (Korean fermented vegetables) as a probiotic food. *J. Med. Food* **2014**, *17*, 6–20. [CrossRef] [PubMed]
37. Islam, M.S.; Choi, H. Antidiabetic effect of Korean traditional Baechu (Chinese cabbage) kimchi in a type 2 diabetes model of rats. *J. Med. Food* **2009**, *12*, 292–297. [CrossRef]
38. Park, J.A.; Tirupathi Pichiah, P.B.; Yu, J.J.; Oh, S.H.; Daily, J.W., 3rd; Cha, Y.S. Anti-obesity effect of kimchi fermented with Weissella koreensis OK1-6 as starter in high-fat diet-induced obese C57BL/6J mice. *J. Appl. Microbiol.* **2012**, *113*, 1507–1516. [CrossRef]
39. Woo, M.; Kim, M.; Noh, J.S.; Park, C.H.; Song, Y.O. Preventative activity of kimchi on high cholesterol diet-induced hepatic damage through regulation of lipid metabolism in LDL receptor knockout mice. *Food Sci. Biotechnol.* **2018**, *27*, 211–218. [CrossRef]
40. An, S.Y.; Lee, M.S.; Jeon, J.Y.; Ha, E.S.; Kim, T.H.; Yoon, J.Y.; Ok, C.O.; Lee, H.K.; Hwang, W.S.; Choe, S.J.; et al. Beneficial effects of fresh and fermented kimchi in prediabetic individuals. *Ann. Nutr. Metab.* **2013**, *63*, 111–119. [CrossRef]
41. Choi, I.H.; Noh, J.S.; Han, J.S.; Kim, H.J.; Han, E.S.; Song, Y.O. Kimchi, a fermented vegetable, improves serum lipid profiles in healthy young adults: Randomized clinical trial. *J. Med. Food* **2013**, *16*, 223–229. [CrossRef]
42. Kim, E.K.; An, S.Y.; Lee, M.S.; Kim, T.H.; Lee, H.K.; Hwang, W.S.; Choe, S.J.; Kim, T.Y.; Han, S.J.; Kim, H.J.; et al. Fermented kimchi reduces body weight and improves metabolic parameters in overweight and obese patients. *Nutr. Res.* **2011**, *31*, 436–443. [CrossRef]
43. Park, J.M.; Lee, W.H.; Seo, H.; Oh, J.Y.; Lee, D.Y.; Kim, S.J.; Hahm, K.B. Fecal microbiota changes with fermented kimchi intake regulated either formation or advancement of colon adenoma. *J. Clin. Biochem. Nutr.* **2021**, *68*, 139–148. [CrossRef] [PubMed]
44. Han, Y.M.; Kang, E.A.; Park, J.M.; Oh, J.Y.; Lee, D.Y.; Choi, S.H.; Hahm, K.B. Dietary intake of fermented kimchi prevented colitis-associated cancer. *J. Clin. Biochem. Nutr.* **2020**, *67*, 263–273. [CrossRef] [PubMed]
45. Park, S.E.; Kwon, S.J.; Cho, K.M.; Seo, S.H.; Kim, E.J.; Unno, T.; Bok, S.H.; Park, D.H.; Son, H.S. Intervention with kimchi microbial community ameliorates obesity by regulating gut microbiota. *J. Microbiol.* **2020**, *58*, 859–867. [CrossRef]
46. Lee, S.H.; Whon, T.W.; Roh, S.W.; Jeon, C.O. Unraveling microbial fermentation features in kimchi: From classical to meta-omics approaches. *Appl. Microbiol. Biotechnol.* **2020**, *104*, 7731–7744. [CrossRef] [PubMed]
47. Oh, M.R.; Jang, H.Y.; Lee, S.Y.; Jung, S.J.; Chae, S.W.; Lee, S.O.; Park, B.H. Lactobacillus plantarum HAC01 Supplementation Improves Glycemic Control in Prediabetic Subjects: A Randomized, Double-Blind, Placebo-Controlled Trial. *Nutrients* **2021**, *13*, 2337. [CrossRef]

48. Youn, H.S.; Kim, J.H.; Lee, J.S.; Yoon, Y.Y.; Choi, S.J.; Lee, J.Y.; Kim, W.; Hwang, K.W. Lactobacillus plantarum Reduces Low-Grade Inflammation and Glucose Levels in a Mouse Model of Chronic Stress and Diabetes. *Infect. Immun.* **2021**, *89*, e0061520. [CrossRef] [PubMed]
49. Park, J.E.; Oh, S.H.; Cha, Y.S. Lactobacillus Brevis OPK-3 from Kimchi Prevents Obesity and Modulates the Expression of Adipogenic and Pro-Inflammatory Genes in Adipose Tissue of Diet-Induced Obese Mice. *Nutrients* **2020**, *12*, 604. [CrossRef]
50. Coppola, S.; Avagliano, C.; Calignano, A.; Berni Canani, R. The Protective Role of Butyrate against Obesity and Obesity-Related Diseases. *Molecules* **2021**, *26*, 682. [CrossRef]
51. Brahe, L.K.; Astrup, A.; Larsen, L.H. Is butyrate the link between diet, intestinal microbiota and obesity-related metabolic diseases? *Obes. Rev.* **2013**, *14*, 950–959. [CrossRef]
52. Zhao, Z.H.; Wang, Z.X.; Zhou, D.; Han, Y.; Ma, F.; Hu, Z.; Xin, F.Z.; Liu, X.L.; Ren, T.Y.; Zhang, F.; et al. Sodium Butyrate Supplementation Inhibits Hepatic Steatosis by Stimulating Liver Kinase B1 and Insulin-Induced Gene. *Cell. Mol. Gastroenterol. Hepatol.* **2021**, *12*, 857–871. [CrossRef]
53. Deng, M.; Qu, F.; Chen, L.; Liu, C.; Zhang, M.; Ren, F.; Guo, H.; Zhang, H.; Ge, S.; Wu, C.; et al. SCFAs alleviated steatosis and inflammation in mice with NASH induced by MCD. *J. Endocrinol.* **2020**, *245*, 425–437. [CrossRef]
54. Oh, J.H.; Lee, J.H.; Cho, M.S.; Kim, H.; Chun, J.; Lee, J.H.; Yoon, Y.; Kang, W. Characterization of Gut Microbiome in Korean Patients with Metabolic Associated Fatty Liver Disease. *Nutrients* **2021**, *13*, 1013. [CrossRef] [PubMed]
55. Zhang, Z.; Zhou, H.; Guan, M.; Zhou, X.; Liang, X.; Lv, Y.; Bai, L.; Zhang, J.; Gong, P.; Liu, T.; et al. Lactobacillus casei YRL577 combined with plant extracts reduce markers of non-alcoholic fatty liver disease in mice. *Br. J. Nutr.* **2021**, *125*, 1081–1091. [CrossRef]
56. Parra-Vargas, M.; Rodriguez-Echevarria, R.; Jimenez-Chillaron, J.C. Nutritional Approaches for the Management of Nonalcoholic Fatty Liver Disease: An Evidence-Based Review. *Nutreints* **2020**, *12*, 3860. [CrossRef]
57. Tendler, D.; Lin, S.; Yancy, W.S.; Mavropoulos, J.; Sylvestre, P.; Rockey, D.C.; Westman, E.C. The Effect of a Low-Carbohydrate, Ketogenic Diet on Nonalcoholic Fatty Liver Disease: A Pilot Study. *Dig. Dis. Sci.* **2007**, *52*, 589–593. [CrossRef] [PubMed]
58. Bueno, N.B.; de Melo, I.S.; de Oliveira, S.L.; da Rocha Ataide, T. Very-low-carbohydrate ketogenic diet v. low-fat diet for long-term weight loss: A meta-analysis of randomised controlled trials. *Br. J. Nutr.* **2013**, *110*, 1178–1187. [CrossRef]
59. Cunha, G.M.; Guzman, G.; Correa De Mello, L.L.; Trein, B.; Spina, L.; Bussade, I.; Marques Prata, J.; Sajoux, I.; Countinho, W. Efficacy of a 2-Month Very Low-Calorie Ketogenic Diet (VLCKD) Compared to a Standard Low-Calorie Diet in Reducing Visceral and Liver Fat Accumulation in Patients With Obesity. *Front. Endocrinol.* **2020**, *11*, 607. [CrossRef]
60. Watanabe, M.; Risi, R.; Camajani, E.; Contini, S.; Persichetti, A.; Tuccinardi, D.; Ernesti, I.; Mariani, S.; Lubrano, C.; Genco, A.; et al. Baseline HOMA IR and Circulating FGF21 Levels Predict NAFLD Improvement in Patients Undergoing a Low Carbohydrate Dietary Intervention for Weight Loss: A Prospective Observational Pilot Study. *Nutrients* **2020**, *12*, 2141. [CrossRef]
61. Pérez-Guisado, J.; Muñoz-Serrano, A. The effect of the Spanish Ketogenic Mediterranean Diet on nonalcoholic fatty liver disease: A pilot study. *J. Med. Food* **2011**, *14*, 677–680. [CrossRef]
62. Luukkonen, P.K.; Dufour, S.; Lyu, K.; Zhang, X.M.; Hakkarainen, A.; Lehtimäki, T.E.; Cline, G.W.; Petersen, K.F.; Shulman, G.I.; Yki-Järvinen, H. Effect of a ketogenic diet on hepatic steatosis and hepatic mitochondrial metabolism in nonalcoholic fatty liver disease. *Proc. Natl. Acad. Sci. USA* **2020**, *117*, 7347–7354. [CrossRef]
63. Mardinoglu, A.; Wu, H.; Bjornson, E.; Zhang, C.; Hakkarainen, A.; Räsänen, S.M.; Lee, S.; Mancina, R.M.; Bergentall, M.; Pietiläinen, K.H.; et al. An Integrated Understanding of the Rapid Metabolic Benefits of a Carbohydrate-Restricted Diet on Hepatic Steatosis in Humans. *Cell Metab.* **2018**, *27*, 559–571.e555. [CrossRef] [PubMed]
64. Anekwe, C.V.; Chandrasekaran, P.; Stanford, F.C. Ketogenic Diet-induced Elevated Cholesterol, Elevated Liver Enzymes and Potential Non-alcoholic Fatty Liver Disease. *Cureus* **2020**, *12*, e6605. [CrossRef] [PubMed]
65. Watanabe, M.; Tuccinardi, D.; Ernesti, I.; Basciani, S.; Mariani, S.; Genco, A.; Manfrini, S.; Lubrano, C.; Gnessi, L. Scientific evidence underlying contraindications to the ketogenic diet: An update. *Obes. Rev.* **2020**, *21*, e13053. [CrossRef] [PubMed]
66. D'Abbondanza, M.; Ministrini, S.; Pucci, G.; Nulli Migliola, E.; Martorelli, E.E.; Gandolfo, V.; Siepi, D.; Lupattelli, G.; Vaudo, G. Very Low-Carbohydrate Ketogenic Diet for the Treatment of Severe Obesity and Associated Non-Alcoholic Fatty Liver Disease: The Role of Sex Differences. *Nutrients* **2020**, *12*, 2748. [CrossRef]
67. Bruschi, F.V.; Claudel, T.; Tardelli, M.; Caligiuri, A.; Stulnig, T.M.; Marra, F.; Trauner, M. The PNPLA3 I148M variant modulates the fibrogenic phenotype of human hepatic stellate cells. *Hepatology* **2017**, *65*, 1875–1890. [CrossRef]
68. Hellemans, K.; Verbuyst, P.; Quartier, E.; Schuit, F.; Rombouts, K.; Chandraratna, R.A.; Schuppan, D.; Geerts, A. Differential modulation of rat hepatic stellate phenotype by natural and synthetic retinoids. *Hepatology* **2004**, *39*, 97–108. [CrossRef]
69. Videla, L.A.; Valenzuela, R. Perspectives in liver redox imbalance: Toxicological and pharmacological aspects underlying iron overloading, nonalcoholic fatty liver disease, and thyroid hormone action. *Biofactors* **2022**, *48*, 400–415. [CrossRef]

MDPI AG
Grosspeteranlage 5
4052 Basel
Switzerland
Tel.: +41 61 683 77 34

Nutrients Editorial Office
E-mail: nutrients@mdpi.com
www.mdpi.com/journal/nutrients

www.ingramcontent.com/pod-product-compliance
Lightning Source LLC
Chambersburg PA
CBHW041953260926
51850CB00019B/2698
9783725847037